Engineering Legends

Related History and Heritage Titles

*America Transformed: Engineering and Technology
in the Nineteenth Century*
Dean Herrin
2003. ISBN 0-7844-0529-8

American Civil Engineering History: The Pioneering Years
Bernard G. Dennis, Jr., Robert Kapsch, Robert J. LoConte,
and Bruce W. Mattheiss (Editors)
2003. ISBN 0-7844-0654-5

Engineering History and Heritage
Jerry R. Rogers (Editor)
1998. ISBN 0-7844-0394-5

*In the Wake of Tacoma: Suspension Bridges and the Quest for
Aerodynamic Stability*
Richard Scott
2001. ISBN 0-7844-0542-5

*International Engineering History and Heritage: Improving Bridges to
ASCE*
Jerry R. Rogers and Augustine Fredrich
2001. ISBN 0-7844-0594-8

Karl Terzaghi: The Engineer as Artist
Richard E. Goodman
1999. ISBN 0-7844-0364-3.

Machu Picchu: A Civil Engineering Marvel
Kenneth R. Wright and Alfredo Valencia Zegarra
2000. ISBN 0-7844-0444-5

Sons of Martha: Civil Engineering Readings in Modern Literature
Augustine Frederich
1989. ISBN 0-8726-2720-9

Engineering Legends

Great American Civil Engineers

32 PROFILES OF INSPIRATION AND ACHIEVEMENT

Richard G. Weingardt
P.E., HON.M.ASCE

Library of Congress Cataloging-in-Publication Data

Weingardt, Richard.
    Engineering legends : great America civil engineers : (32 profiles of inspiration and achievement) / Richard Weingardt.
        p.   cm.
    Includes bibliographical references and index.
    ISBN 0-7844-0801-7
    1. Civil engineers—United States—Biography. I. Title: Great American civil engineers, 32 profiles of inspiration and achievement. II. Title: American civil engineers, 32 profiles of inspiration and achievement. III. Title: Civil engineers, 32 profiles of inspiration and achivement. IV. Title.
    TA139W35   2005
    624'.092'273—dc22   2005014120

Published by American Society of Civil Engineers
1801 Alexander Bell Drive
Reston, Virginia 20191-4400
www.pubs.asce.org

To my wife, Evelyn

Contents

List of Illustrations

Foreword

I first encountered Richard Weingardt's writing in the magazine *Structural Engineer*, to which he contributes his regular column, "The View from Here," and I immediately became a fan. Each month, he explores a different aspect of engineering and reminds us how proud we are to be members of such a creative and essential profession. His eclectic interests range from our profession's historical roots to its technological future, and he promotes a strong sense of continuity across generations of engineers and their works. Although the tools of today's engineers may be different from those of centuries past, Rich Weingardt constantly reminds us that the aspirations of engineers remain solidly grounded in the tradition of serving society and advancing civilization.

Rich Weingardt is a conscience of—and a cheerleader for—the engineering profession. But he is no idealistic zealot. Like engineers in general, he is a realist. Also, as an engineer, he recognizes that the profession is not as efficient as he imagines it could be, and he wants to improve it. But engineers do not just wish for a better world; they do something to make it better. Whenever Rich Weingardt writes about a fault, he also puts forth a remedy. If engineers think that they and their profession are not getting respect and recognition from the public and their surrogates, the ubiquitous news media, then Rich offers a proposal to improve the situation—specifically to take every opportunity to explain and promote the profession to the public and the media in words and deeds they will notice and understand. Rich Weingardt has certainly done more than his share in this regard.

In addition to his column in *Structural Engineer*, Rich also writes the "Engineering Legends" column in the journal *Leadership and Management in Engineering*, which is published by the American Society of Civil Engineers. It is from that column that this book takes its title. And it is from five years' of columns that it takes its form of biographical exposition.

Every profession begins with its practitioners, and engineering is no exception. To understand the essence of this profession fully, we must understand its founders and its pioneers, their intentions and their efforts, their achievements and their dreams. Rich has a broad knowledge and understanding of these things as well as an unmistakable talent for conveying them with enthusiasm to his fellow engineers and to the public.

The technology of yesteryear may be past technology, but its history still holds great lessons for us about the nature of engineering. Similarly, the biographies of engineers of the past can provide us with considerable insight into the soul of the engineer. *Engineering Legends* is not a history of engineering, but rather a series of stories about outstanding engineers whose names and lives should be known by all their professional descendants and by the public as well.

Samuel Smiles, in his monumental nineteenth-century series of biographies, *Lives of the Engineers,* presented them not only as the shapers of the Industrial Revolution but also as moral individuals whose lives were paradigms of personal achievement. His *Lives* did much to establish the canon of Victorian British engineers that we know today, embodied in the likes of James Watt, John Smeaton, Thomas Telford, George and Robert Stephenson, and Isambard Kingdom Brunel. Like Smiles in his *Lives*, Rich Weingardt in his *Engineering Legends* sees the value of explicating the personal development of engineers. Even if at the expense of delving deeply into their technical achievements, Rich presents a tentative pantheon of American civil engineers whose lives and careers he regards as exemplary.

Among the characteristics of engineers that come out so vividly in Rich Weingardt's sketches of their lives are breadth and variety. Although Rich is a structural engineer, he has chosen his exemplars from far afield, ranging from environmental experts to transportation trendsetters. But the mistaken stereotype of the engineer as an uncultured, antisocial, uncommunicative, technical automaton is nowhere to be found in *Engineering Legends*. Rather, we find professionals who have an enormous range of experience and achievement outside their primary profession. We find that Kate Gleason, who pioneered mass-produced and low-cost concrete housing, also had a career as a banker and that the environmental engineer Ellen Swallow Richards cofounded what is now the American Association of University Women. We encounter bridge builders as artists in more than just concrete and steel: David Steinman as an accomplished poet and Ralph Modjeski as a talented and dedicated pianist. We also learn that the structural engineer Willard Simpson was a civic leader in his native San Antonio; that the one-time atheist Fred Severud became a fervent Christian who preached and taught the Bible; and that T. Y. Lin, who was known as Mr. Prestressed Concrete, had a passion for ballroom dancing.

Each chapter of *Engineering Legends* has little surprises like these, which add unexpected dimensions to the professional résumés of the engineers Rich writes about. His book conveys the distinct impression that he revels in discovering and communicating the full lives that engineers lead to everyone whose attention he can attract. Around biographical nuggets, he constructs essays that collectively reveal the engineer as hero and model, and he portrays the profession as a rewarding and satisfying one to individuals from all walks of life and of every inclination.

Rich Weingardt has done the engineering profession a great service in writing and collecting these essays into a book that cannot help but inspire engineers, future engineers, and all who benefit (and will continue to benefit) from their work. *Engineering Legends* is a book that has something for everyone.

HENRY PETROSKI, P.E., F.ASCE
Aleksandar S. Vesic Professor of Civil Engineering
and Professor of History
Duke University
Chair, ASCE History and Heritage Committee

Preface

With its more than 10 million practitioners around the world, engineering is the second largest profession worldwide, following teaching. Within its ranks are many superstars and trendsetters. Nearly one-fourth of the world's engineers currently work and/or reside in the United States, and more than its share of outstanding engineers are among those in the pantheon of civil engineering legends. In this book, 32 among the "best of the best" are spotlighted, and many others are cited. In total, these mentors and role models have helped define the profession as it is today.

Engineering—civil engineering in particular—is the single most critical group needed for maintaining any nation's infrastructure and standard of living, the United States included. Yet, the general public and media remain mostly unaware and uninformed about engineering accomplishments. They do not know what engineers actually do and who they are because engineering accomplishments are rarely headlined in the news. In addition, few general-readership books and full-length articles have ever featured civil engineers and what makes them tick.

Engineering Legends seeks to correct that. It is the first twenty-first century book devoted to introducing a broad cross section of civil engineering pacesetters, from the late 1700s to the early twenty-first century, who have quietly helped shape our nation and communities. Its collection of biographies describing broad-visioned engineering greats—and their feats—illustrates why the history of U.S. civil engineering is synonymous with the history of progress and civilization. And history, when all is said and done, is about people. Said Ralph Emerson, "There is properly no history; only biography."

For practicing engineers, young and old alike, this book is a means for discovering more about their noble profession—and about its history and those who made it what it is today. The biographies of the featured legends will inspire readers—both engineers and nonengineers—to reach for their own greatness. Said William Moore, cofounder of the international engineering firm of Dames and Moore, "Understanding what went before us and having more knowledge of what was accomplished by previous generations makes us better engineers, a lot prouder of the profession and ourselves for being in it."

The field of civil engineering encompasses a broad array of specialties and subspecialties, including expertise in transportation, water resources, sewerage, municipal planning, environmental services, surveying, and geotechnical and structural design. *Engineering Legends* features noteworthy individuals from every corner of the country and social stratum in society who not only excelled in these activities but also influenced the direction taken by both their communities and the civil engineering industry. It depicts civil engineers—from consulting, industry, academia, and government—as individuals who readers would find exciting to know. Through exposure to their examples and words of advice, they can learn how to succeed both as professionals and as human beings.

The legends selected are all U.S. civil engineers who made their marks as Americans (even though a number were born and educated overseas). If not deceased, they are in their twilight years or fully retired from engineering. Not all are well known within technical or academic circles. But because of what they did, how they did it, and their courage in overcoming obstacles, all have helped shape history and have significantly impacted civil engineering practices in this country and beyond. The desire to convey that this profession has attracted a full range of multidimensional role models and colorful mentors diverse in gender and ethnic origin influenced the selection of those featured. That the legends came from various parts of the country was also a consideration.

This book reveals private facts about these legends: their parents and siblings, growing-up years, college days, marriages and family influences, and career challenges and triumphs. Included are their anecdotes and words of wisdom, plus their pithy comments on such issues as why they became civil engineers, what is right or wrong about the profession, and how it can be improved.

Because younger Americans today want to know about the heroes and heroines in the vocation they have chosen (or will pursue), they now have a wide-reaching reference to consult—*Engineering Legends*. It reveals to them that many types of superstars dwell within the ranks of civil engineering; that many have gone beyond being narrowly focused technical wizards; that civil engineering leaders were (and are) involved in shaping this nation's future; and that civil engineering is, indeed, relevant to progress, the economy, and America's standard of living.

Engineering Legends has been written for practicing civil engineers (both young and seasoned), engineering professors, college students, potential engineers (youngsters currently in junior high and high school), school counselors, the general public, and the media that inform these groups. Indeed, readers include all those wanting to know more about civil engineering and/or engineers, what they have accomplished, and how they think.

The book would make a perfect gift for an engineer who is well entrenched in the profession, just starting out, or in mid-career. It is also ideal for anyone as a birthday, Christmas, or graduation present and would serve public libraries as well as junior high and high schools.

Engineering Legends features more than 60 photographs, including one photo of every engineering legend and another of one of his or her notable achievements. Photos of these projects—like the familiar Sears Tower, St. Louis Arch, and Brooklyn Bridge—convey the wonder and practicality of civil engineering works all around us. Additionally, the book integrates the history of engineering into the general history of the United States and discusses engineering accomplishments, *not* as if they occurred in a historic void but as one of many social enterprises that built this country.

To this end, the book is broken into eight chapters of legends, four legends per chapter. Lead-in remarks for each chapter (for Chapters 1 to 8) put into context how the civil engineering disciplines of the featured engineers fit into the profession as a whole as well as the time frames in which they lived. These lead-ins, as well as the book's Introduction, are written to be of appeal to nonengineers and engineers alike.

The final chapter (Chapter 9) speaks to the future, suggests who will be tomorrow's engineering legends, and lays out what the civil engineering community must do to be a viable force in the coming years.

A number of approaches to using the book are possible, the most obvious of which is to consecutively read from beginning to end, first the Introduction and then all nine chapters, in one or more sittings. Another is to read the chapters in any order according to the reader's interest in the branch of civil engineering or phase of history each highlights. Or, because each of the biographies are stand-alone essays, a third way is to select and read one biography at a time, in any order.

Readers who want to learn more are invited to refer to the materials listed in the References section. In addition to those sources, books and articles are occasionally being written about—and/or updated on—civil engineering greats, including many not profiled or cited in *Engineering Legends*. Browsing the Internet via search engines like Google or Yahoo can uncover real gems.

Legendary men and women leave great legacies and inspire greatness in others, young and old alike. Legends in the field of civil engineering—whether mentioned in this book or not—are no exception. Readers are encouraged to produce and circulate similar biographies featuring great civil engineers in their own locale. Their stories simply need to be told.

Acknowledgments

It is amazing where many priceless facts and anecdotes come from. Yes, they come from traditional sources—books, articles, encyclopedias, newspapers and magazines, obituaries, personal papers and letters, and so on—but many times they come from oral histories privately recorded by those who knew (or knew of) the person personally. This gathering of relevant information occurs over a substantial amount of time. After a while, a certain amount of dust accumulates and some names simply fade from memory.

Similarly, many people over the years have contributed greatly to this project whether they knew it or not. And although their names are not specifically mentioned, I nevertheless thank them sincerely for their help and assistance. I appreciate their insights, knowledge, and experiences; I thank them for their encouragement and enthusiasm about this project. Additionally, I want to acknowledge the countless engineers who, over the years, responded to my numerous surveys to identify legends, role models, and mentors in the field of civil engineering.

I am also indebted to the following special people for providing noteworthy anecdotes and facts, helping gather information and photographs, and/or offering advice on how to locate and identify worthy legends: Osman Akan, Mir Ali, William Andrews, Dinah Everett, Larry Feeser, David Fowler, Jan Gleason, William Hall, David Hargrove, Robert Johnson, William Moore, Jr., Lee Overcamp, Jan Plachta, James Poirot, David Ruby, Jeffery Russell, Fred Severud, Jr., Jay Simpson, Robert Sinn, Douglas Steadman, Stein Sture, Jane Tiedeman, Jill Tietjen, James Yao, and John Zils.

Many of this book's biographies have appeared in one form or another in my column "Engineering Legends" in the American Society of Civil Engineers' *Leadership and Management in Engineering* (LME) journal from 1999 to 2004. I am indebted to the ASCE Journals Department for allowing me to reuse them here. I very much thank Henry Petroski, America's Poet Laureate of Technology, for composing the foreword, and Barbara McNichol for her outstanding assistance with editing.

A time-consuming endeavor such as this could not have been done without the full support and encouragement of my office staff and family members. I am deeply thankful to each of them—especially the president of my company, John Davis, and my wife, Evie, and our children Nancy, Susan (and her husband George), and David (and his wife Kathy).

RICHARD G. WEINGARDT, P.E., HON.M.ASCE

Introduction

*He who knows only his own generation
remains always a child.*
—George Norlin (inscription engraved
in granite over the entrance to the
University of Colorado Library)

Many people today perceive that engineers make things run but they do *not* run things. This book thoroughly debunks that perception. The lives and actions of the 32 engineering legends spotlighted, and many others mentioned, fully dispel the misconception that civil engineers are not societal leaders capable of heading up significant nonengineering ventures.

Even though few in the media recognize and accentuate engineering contributions, the legends profiled in this book confirm that, throughout the ages, the role played by civil engineers has been crucial for advancing civilizations. Engineers have provided citizens of nations with the necessary infrastructure for creating wealth and contributing to progress over the millennia.

According to Richard Kirby et al., in *Engineering in History*, "An inescapable conclusion to be drawn from the story of engineering in history is that engineering has become an increasingly powerful factor in the development of civilization." Engineering, the historians reported, "does not occur in a historical vacuum without reference to other human activities," and the fundamental changes stimulated by civil engineering developments "accelerate the rate of historical and social revolution" (Kirby 1956).

As the single most indispensable group needed for developing and maintaining a nation's built environment and standard of living, civil engineers play an extremely large role on the world's stage and in the daily lives of people. These pragmatic problem solvers protect the public's health, welfare, and safety by giving it clean water and air, modern transportation, and safe and sound shelter. It would be impossible to imagine modern life without the many contributions of civil engineers. Indeed, they are the stewards of infrastructure worldwide.

Dream Profession

If you had to invent a dream profession with notable characters in it, you could not come up with anything better than civil engineering. What engineers do adds value. Many visionaries and think tankers consider engineers the *wealth creators*, instrumental in helping enlarge the economic pie for everyone.

Even though they possess worthy credentials and deserve high accolades, engineers generally remain a humble bunch; they do not like to brag about their accomplishments. For that reason, engineers themselves inadvertently contribute to the "invisible profession" syndrome they find themselves in today. Many more books like *Engineering Legends* that are generated for consumption by nonengineers are needed to change this.

Plus, more must be made of the fact that famous Renaissance men of the distant past were civil engineers—the greatest of them all, Leonardo da Vinci (1452–1519). A celebrated Italian artist and inventor, da Vinci was also a scientist and civil engineer. For example, he designed an ambitious and magnificent bridge to cross the mighty Bosporus River at Istanbul, Turkey, although it was never built. However, a smaller-scale, but equally magnificent, version of his design was completed near the tiny town of As in southern Norway in 2001 (BBC 2001).

In 1482, the most famous Renaissance man of all time was employed by the Duke of Milan as the "painter and engineer of the duke." Da Vinci gave advice on cathedrals and other construction projects, and he became heavily involved with hydraulic engineering. He also devised successful plans for draining vast sections of Italian countryside and for building canals, including the Adda Canal connecting Milan to Lake Como. His well-known notebooks reveal his passion for things mechanical, ranging from complex construction cranes and drilling machines to underwater breathing equipment and flying devices. During a number of wars within Italy, da Vinci also served as a military engineer and created assault machines, armaments, and pontoons, and even a steam cannon (Sekine-Pettite 1999).

Definition of Civil Engineering

The standard definition of most of today's State Boards of Registration for professional engineers (PEs) is:

> Civil engineering embraces all studies, analyses and activities, including design and construction, in connection with fixed works of irrigation, drainage, waterpower, water supply, flood control, inland waterways, harbors, municipal improvements, railroads, highways, tunnels, airports and airways, purification of water, sewerage, refuse disposal, foundations, grading, framed and homogeneous structures, buildings, or bridges.
>
> Civil engineers provide construction project management services as well. They also locate and establish the alignment and/or elevation for any

of the fixed works embraced within the practice of civil engineering. Plus, they determine the configuration and/or contour of the earth's surface and the position of fixed objects thereon or related thereto.

Prior to and at the beginning of the twentieth century, few women were graduate civil engineers and involved in the profession. Elsie Eaves (1898–1983) was a notable exception. After graduating from the University of Colorado in 1920, she embarked on an enormously illustrious career beginning with the U.S. Bureau of Public Roads. She went on to work with Denver and Rio Grande Railroad, Colorado Highway Department; Herbert Crocker, a prominent consulting engineer in Denver; and Crocker and Fisher Construction, also in Denver. Then she headed east and worked at McGraw-Hill in New York City.

In a publication focusing on interesting women in engineering as a career, *An Outline of Careers for Women: A Practical Guide to Achievement*, Eaves came up with a less formal and more refreshing definition for the profession. She wrote: "Civil engineering is the broad term that defines the work of reorganizing natural or existing conditions or forces into structures more useful to man's needs. The history of development of modern social and economic life could be written by recording the development of civil engineering—railroad location and construction, good roads, structural bridges, and buildings of present-day magnitude, waters supplies unfailing under the demands of cities dependent upon them" (Fleischman 1929).

Eaves, the first woman licensed as a PE in the state of New York, was named manager of the Business News Department of McGraw-Hill's *Engineering News-Record* (ENR) in 1932. While there, she organized and directed ENR's first measurement of Post War Planning by the Construction Engineering in 1945, which was used both by the government's Committee for Economic Development and by the American Society of Civil Engineers (ASCE) as their official industry progress report.

Eaves had many firsts, among them being the first woman to become a member of ASCE (1927), receive the American Association of Civil Engineers' Award of Merit (1967), earn the University of Colorado's Distinguished Engineer Alumna Award (1973), be honored with the prestigious Norlin Medal (1974) from the University of Colorado, and become an ASCE Honorary Member (1979).

ASCE Infrastructure Report Card

The main categories of work civil engineers are responsible for are outlined in the Infrastructure Report Cards that the ASCE regularly produces on the state of the nation's built environment. The 12 infrastructure categories covered by the Report Cards are roads, bridges, transit, aviation, schools, drinking water, wastewater, dams, solid waste, hazardous waste, navigable waterways, and energy. They represent not only significant areas of work for civil engineers, but also areas of expertise for which they should provide public policy input.

In 2003, ASCE's Report Card gave the U.S. infrastructure a dismal grade of D+ overall. That represents deterioration from a grade of C in 1988, when the first national Infrastructure Report Card was completed by a special commission appointed by President George H. W. Bush. Those findings hardly endorse past policy decisions as being optimum, but they do emphasize the need for more civil engineers to become involved in setting public priorities and direction.

The importance of engineers becoming involved in such activities was also underscored by Bernstein and Lemer in *Solving the Innovation Puzzle: Challenges Facing the U.S. Design and Construction Industry* (1996). They wrote, "The U.S. today possesses a physical infrastructure of extraordinary scale and scope. This civil infrastructure supports virtually all elements of our society, and the people and businesses that have produced it comprise a major segment of our economy. History indicates that the growth, flourishing and decline of any civilization are closely mirrored by the life cycle and performance of its civil infrastructure" (Bernstein 1996).

In the final analysis, how favorably the public perceives civil engineers—and their profession—will tremendously influence how effective they perform as societal leaders and even as technical experts. It will also influence whether civil engineers are thought of as professionals or technicians whose services are treated like a commodity and hired by lowest bid without regard to qualifications.

By identifying and celebrating past accomplishments of top civil engineers, this book aims to encourage more involvement in these issues by today's practicing engineers. When that happens, the public will begin to see the need to have increasing numbers of civil engineering leaders running things as well as making them run.

Engineering Legends

The biographies of those featured in this book reveal exceptional profiles of courage from daring and visionary men and women whose defining moments have set the compass for the profession—and helped the country move forward, in many cases, in quantum leaps. Throughout the book, their lives of inspiration and achievement are put into a historic context.

Of the many engineers studied, those selected for this book represent the broadest range possible from all stations in life, as well as ethnic and gender diversity. They come from all parts of the United States and illustrate that anyone with sound vision and intelligence who puts his or her mind to it, no matter the odds, can become a legend in life and in civil engineering or any related profession.

In any publication profiling only 30 or so engineering superstars, it would be impossible to include every noteworthy civil engineer and come up with a list with which everyone agrees. Although the legends featured here represent many of the best of the best from the wide range of disci-plines in civil engineering, they are only a fraction of the "unsung heroes and heroines" who have graced the pages of civil engineering history. Many more could be included if space allowed.

Some prominent and historically significant civil engineers such as Karl Terzaghi (the father of soils engineering), John Stevens and George Goethals (of Panama Canal fame), and Othmar Ammann (designer of the Verrazano Narrows Bridge) have been adequately covered in previous publications (several are listed in the References) and are not profiled in depth here. In *Engineering Legends*, they, and others like them, however, are mentioned in passing either in the introductory (lead-in) remarks to the chapter featuring their discipline or within the story of a legend if they were a colleague or peer of that person.

The first eight chapters contain the biographies of the book's selected American civil engineering greats. The first four chapters are the most his-toric, showcasing pathfinders, canal makers, railroaders, tall building pio-neers, environmental experts, and transportation trailblazers. Chapters Five and Six illuminate daring engineering innovators, mostly in the area of structural wonders. Chapter Seven spotlights the movers and shakers of the industry, and Chapter Eight reflects on civil engineering educators extraor-dinaire. Chapter Nine looks to the future.

For sure, *Engineering Legends* is informative, but it is also intended to spark reflection and insight into what happened in the past and what the reader can do to alter the future. It is meant not as an ending but as a begin-ning in the celebration of civil engineering masters. Readers are encouraged to spread the word, to step forward and be heard—and to come up with their own list of noteworthy legends and produce similar publications.

CHAPTER ONE

Empire Makers

Bring me men to match my mountains,
Bring me men to match my plains,
Men with empires in their purpose
And new eras in their brains.
—Sam Walter Foss

From day one of the exploration and development of the United States, civil engineering played a major role. First came the need for addressing the requirements of housing and shelter, food production and storage, sanitation and water supply, and the development of docks and harbors to maintain a link to the settlers' countries of origination—and known civilization. Meeting these needs demanded an ever-increasing application of pragmatic engineering and technological principles and solutions.

As frontier villages and farms multiplied, it became crucial for colonial America to develop reliable roads to connect neighboring villages and centers of commerce. With this came the desire to explore, survey, and plat more and more of the vast unknown. Some of the expanding country's most educated and one-day-to-be-prominent national leaders—among them, Presidents George Washington and Thomas Jefferson (and later Abraham Lincoln)—spent their early adult years surveying land and doing rudimentary civil engineering.

The phrase *civil engineering* was first introduced and popularized by Bristish-born John Smeaton (1724–1792). He adopted the term *civil engineer* to distinguish himself from the military engineers whose numbers greatly surpassed those of the civilian set. In 1771, five years before the signing of the American Declaration of Independence, Smeaton founded the Society of Civil Engineers (renamed the Smeatonian Society after his death), the Western world's first professional engineering society.

Remembered most for rebuilding the Eddystone Lighthouse (a structure still standing in Britain), Smeaton introduced technical engineering innovations that were universally adopted by engineers everywhere, including early America. His work on waterwheels, for instance, proved that overshot wheel power was twice as efficient as the traditional undershot waterwheel power. This was promptly put to profitable application in colonial milling operations (Sekine-Pettite 1999).

In addition to refining the use of waterpower, early America's expanding population and economy relied on transportation via rivers and eventually canals. Once it was demonstrated that four horses could pull a 1-ton wagonload only 12 miles a day over an ordinary road and 18 miles over a good turnpike, but 100 tons over 24 miles could be hauled by waterway in the same period, the demand for building networks of canals intensified.

The speed and economy of moving goods by canal travel increased immediately after the Revolutionary War (1775–1783), and along with it came a strong trend to expand westward into sparsely settled virgin lands. Using natural and artificial waterways rather than overland trails to move people and their possessions clearly garnered considerable popular postwar support.

That there was a major "canal-building craze" going on in England and much of Europe around the turn of the century (late 1700s to early 1800s) added to the appeal. Many powers that be in several U.S. states were bitten by "canal fever," projecting that building great networks of canals would boost their state economies. However, there were no highly experienced canal engineers or builders in the United States at the time, so all eyes turned back to the homelands of their ancestors for assistance.

The first experienced canal engineers to arrive in America came from England and Holland, where knowledge of building canals was ages old. Most noteworthy of these European experts were England's William Weston and Benjamin Latrobe and Holland's John Christian Senf. By helping transfer the latest canal technology to the United States, all three tremendously influenced America's first generation of canal-building engineers and the success of the new country's economic growth and its emergence as an influential nation.

The first sizable American canal was built in South Carolina in 1786 to connect the Santee and Cooper Rivers above Charleston. Its design and construction was entrusted to Senf, who was eventually named the state's chief engineer for all its construction projects. Other states like Ohio, Pennsylvania, and Massachusetts quickly commissioned canal projects, as did New York, although cautiously at first.

Then in the early 1800s, shortly after President Thomas Jefferson consummated the Louisiana Purchase (1803), New Yorkers began thinking big and started making "men to match my mountains" plans to build the greatest canal system of all—the Erie Canal. This project would rocket forward both the quantity and the quality of America's engineering community forever, and it would spawn countless pioneering civil engineering notables, including industry giants like Benjamin Wright, whose life is highlighted next in this chapter.

After President Jefferson doubled the size of the United States by securing the Louisiana Purchase territories from Napoleon Bonaparte and France, he sent out a team of pathfinders to locate the Northwest Passage to the Pacific Ocean, again underscoring the nation's preoccupation with waterways being the means for the future. Additionally, he commanded the Meriwether Lewis and William Clark–led group to explore and map their trail into the unknown and identify any new fauna and flora along the way. When the Lewis and Clark "Corps of Discovery" finally returned in 1806 with their findings, Lewis and Clark themselves were hailed as national heroes, even though they did not discover a Northwest Passage (Ambrose 1996).

Quite disappointing were the Corps of Discovery's reports that did not offer support for Jefferson's hope that the new territory would provide endless farming land. Lewis reported that the lands were too dry. Plus, he stated that the indigenous Plains Indians were hostile enough that they would most likely block major settlements and trade along the upper Missouri River.

About the time of the Corps's return in 1806, Captain Zebulon Pike was put in charge of a small U.S. army platoon of 23 men. He led them into Colorado to establish peaceful relations with the Indians and map the southern region of the Louisiana Territory. Pike's report, which was published in 1810 during President James Madison's first term, rekindled interest in the west, both in the United States and in Europe. But not much else happened. Further extensive government-sponsored exploration of the area was put on the back burner and would have to wait its day (Stone 1956). That day came 10 years later when a brilliant young civil engineer, Major Stephen Long, headed an expedition back into the Colorado region. (In this book, a description of Long's adventure follows Benjamin Wright's story.)

America's canal-building mania of the early nineteenth century gave way to railroad-building mania in the middle part of the 1800s. Between 1840 and 1850, the country's number of railroad miles increased from 2,818 to 9,021. From 1850 to 1860, they increased from 9,021 to 30,636—more than 340 percent (Griggs 2003). In the midst of this sweeping railroad fever would rise an unassuming young civil engineer of slightly less than average height, Theodore Judah, who would play a pivotal role in advancing railroads both on the East Coast and in America's Wild West.

Judah's undaunted efforts ensuring that the nation's transcontinental rail-line would become reality placed him on a pedestal among the tallest involved in its completion. Although most railway work in the country was sidelined by the Civil War (1861–1865), four years after its bloody conclusion, the United States sported 40 percent of all railroads in the world *and* its first transcontinental rail system. No small thanks were due to men like Judah.

In addition to a massive railroad construction program right after the Civil War, its horrific casualties, both in lives and in property, forced a battered nation to begin rebuilding everywhere and get on with making itself whole again. The era was punctuated with new building projects in almost every part of the country, from coast to coast and north to south. Many civil engineers who served in the war on both sides became actively involved in many of the large-scale construction projects.

In less than two decades, America's cities and built environment reached a size and enormity—and daring—to attract world attention. Of particular note was engineering taller and taller buildings that were redefining skylines in major cities. The final civil engineer featured in this section—William Jenney—has forever been immortalized as the designer of the world's first skyscraper. With the completion of his Home Insurance Building in Chicago in 1885, the new race to build the tallest skyscrapers—either for national pride or for doing your neighbor one better—commenced.

Beginning with the Home, the United States became the recognized global leader in high-rise construction and stayed so for many years with notable record holders like the Empire State Building and Sears Tower. Today, however, the race to build the tallest building is being run in Asia and the Middle East. Americans have taken this attitude: "Been there, done that." Having the tallest building is no longer as compelling for Americans as it is for people trying to show that their country's economy has "arrived" (Iovine 2004).

Although this nation's early development witnessed many significant history-altering phases (and countless civil engineering greats involved in the success of each), only four are highlighted in this section: (1) canal building, (2) opening the west for expansion, (3) developing a nationwide railroad system, and (4) rebuilding cities after the calamitous Civil War. Likewise, only four great engineers are featured in this chapter. All were daring empire-making pioneers, civil engineers whose names will be emblazoned onto the pages of history for all time—Wright, Long, Judah, and Jenney.

Benjamin Wright

Declared the "Father of American Civil Engineering" by the American Society of Civil Engineers (ASCE) in 1969, Benjamin Wright had an enormously colorful life and far-reaching influence. Active on the American scene right after the nation's birth—and during the heyday of the Industrial Revolution—he made his mark as a pioneer engineer, legislator, judge, businessman, and community pacesetter. His lifelong pursuits and achievements qualify him for history's pantheon of noteworthy Renaissance men.

From the late 1700s through the first half of the 1800s, Wright was a major player in developing and building American's transportation and canal systems, especially the Erie Canal network, which was the envy of the world by the early part of the nineteenth century. He was the mentor to—and came to be called the "professor" by—young on-the-job-educated engineers who earned their engineering stripes working with him.

Wright's history-altering Erie project—with its more than 365 miles of canals and a 555-foot descent from Lake Erie——served as the training ground and the first practical school of engineering in the United States. By the time the project was opened for business in 1827, 57-year-old "professor" Wright had played a major role in preparing hundreds of America's brightest youngsters for careers as civil engineers. They in turn educated countless others in the profession. By the mid-1800s, almost every civil engineer in the United States, it seemed, had trained with (or been trained by) Wright or someone who had worked under him on the Erie Canal.

Benjamin was born to Ebenezer and Grace (Butler) Wright in Wethersfield, Connecticut, on October 10, 1770, shortly before the beginning of the American Revolution. Although his father was a prominent lieutenant in George Washington's Continentals, he remained debt-ridden long after the conclusion of the Revolutionary War. And his children received limited formal schooling. Fortunately for young Benjamin, his much-better-off uncle Joseph took him under his wing and saw to it that he got a solid education in math as well as the basics of surveying and law.

Benjamin Wright

Photo credit: Milstein Division of U.S. History, Local History & Genealogy, The New York Public Library, Astor, Lenox and Tilden Foundations

In 1789, when his parents and younger siblings moved to the township of Rome in upstate New York, where his father tried his hand at farming, 19-year-old Benjamin followed along. However, he did not go into tilling the soil; rather, he took up surveying. Surveyors were in hot demand in the newly opened wilderness in central New York, so he did not have to look far for work. He quickly developed an excellent reputation for accuracy, honesty, and reliability, and with it came ever-increasing calls for his services. In short order, he had as much business as he could handle, and, within a few years, he had surveyed hundreds of thousands of acres of virgin land in the Oneida Lake district.

In 1794, when Benjamin was 24 years old, he was hired by the famous English civil engineer William Weston to assist him in making canal surveys and plans for a collection of canals that would eventually become part of the Erie Canal network. Shortly after this engagement, the up-and-coming young Wright was elected to the New York State Legislature, where he cut his eyeteeth concerning effective leadership in the public arena.

The British-born Weston had engineered a number of important canals, roads, and bridges in Europe. He initially made his presence felt in America in Pennsylvania, where he supervised the building of the groundbreaking canal between the Schuylkill and Susquehanna Rivers. Along with his up-to-date engineering skills, Weston brought with him a sophisticated, optical surveying level unknown in America at the time—the Troughton "Wye Level." The Troughton instruments were soon being used on every canal project in the United States, and Weston quickly became the catalyst for a proliferation of a whole new generation of Americans—skilled canal surveyors, builders, and engineers.

After Philadelphia, Weston moved to Boston to complete the Middlesex Canal from Boston to Lowell. It had 20 locks, seven aqueducts, and 50 bridges, and was the field-study project for the budding U.S. engineers who would work on the great Erie Canal project.

In the mid-1790s, New Yorkers were eager to get into the canal transportation business to compete with Pennsylvania and Ohio for the riches the nation's burgeoning western expansion promised. The state hired Weston full-time to make surveys and develop designs for canal projects. With young Wright once more as his assistant, Weston laid out several canals and locks on the Mohawk River, including a major link connecting the Mohawk to Wood Creek.

By 1798, several Weston- and-Wright-designed canals were in full operation. These canals allowed large 16-ton Durham boats to move up and down the Mohawk. However, they could not yet reach Lake Ontario or even Albany, the state's capital. That major chapter of American canal building would have to wait nearly two decades until adequate funding could be secured—and the public's will to move forward was also acquired.

On September 27, 1798, Benjamin married Philomela Waterman. They would have nine children, five of whom would follow in their father's footsteps and become civil engineers.

> "*Wright and his assistants built the longest canal in the world, in the least time, with the least experience, for the least money, and the greatest public benefit.*"
>
> NEW YORK STATE ARCHIVES

By 1800, Wright's mentor, Weston, was back in England, and the newly married Wright, on his own again, landed a major commission by the Western Inland Lock and Navigation Company to survey the Mohawk River basin from Schenectady to Rome. The maps he would produce of the rugged region would serve as the master guide for all future development along the waterway. After laying out a practical canal route from Rome (on the Mohawk) to Waterford (on the Hudson) for the New York State Canal Commission, Wright won appointment as a New York county judge. It was a post that would earn him the lifelong moniker "Judge Wright."

By 1813, New York had amassed enough money to commit to final plans for expanding its canal system all the way to Lake Erie. Weston, who by then was entrenched in England, was contacted to review updated construction designs by mail and was offered the prestigious job as chief engineer for the entire Erie project at an enormous salary. The United States and Great Britain, however, were still in the throes of the War of 1812 and Weston refused. The building of the most complex artificial waterway in the world—the greatest engineering feat of its time—would therefore fall entirely on the shoulders of American engineers. And under Judge Wright's able leadership, they would gallantly rise to the occasion.

With funding for the project firmly in place by 1816, the Erie was ready to forge ahead full blast, and 46-year-old Benjamin Wright was officially named the project's chief engineer in 1817. Calling on all his organizational and innate engineering skills—and political adroitness—he molded a group of "Yankee Doodle Dandy–type" workers into a finely tuned team that overcame, with ingenuity, all the countless technical and political difficulties they encountered. Construction was accomplished using horses, mules, wagons, wheelbarrows, hand tools, and thousands of

unproven but hardworking laborers, who were mostly of Irish descent.

When completed in 1827, the Erie Canal revolutionized transport between the eastern and western parts of the Union. Its success spurred America's "canal fever" of the 1820s and 1830s, which was so important to the country's growth and emerging quest for global recognition. In Wright's honor, the first ship to travel the canal system was named *Chief Engineer*. And the demand for individuals from his group of trained canal engineers exploded. They were needed everywhere and ended up developing much of the country's ever-spiraling transportation and infrastructure systems. They defined the engineering profession of America at the time.

Erie Canal Lock 2 in Waterford, New York, 2003

After being in continuous operation for 180-plus years, the Erie system remains remarkably functional. Extensive lock rehabilitation and modernization work at Waterford recently received a prestigious Engineering Excellence Award from ACEC-New York for engineering innovation.

Photo credit: NYS Canal Corporation

Two of Wright's most steadfast Erie aids were James Geddes (1763–1838), who, like Benjamin, was self-taught, older, and seasoned, and youthful Canvass White (1790–1834), who traveled overseas in 1817 to study Europe's canals, locks, dams, and bridges. He also researched the nuances of Europe's popular Portland cement for underwater applications. After a year of reviewing and recording the engineering features of Great Britain's canals, White returned with a portfolio stuffed with copious notes and carefully rendered drawings, and a better knowledge of canal construction—especially locks—than any person in America. He also brought with him the latest surveying instruments.

Armed with his newly acquired knowledge, White became instrumental in raising the Erie engineering team's ability to design and build the latest state-of-the-art waterways, locks, dams, and bridges—designs that were widely copied on subsequent projects. Another major contribution was his development of a waterproof, hydraulic cement (created from native deposits of New England limestone). It, rather than England's Portland cement, became the cement of choice for the concrete portions of the project.

Over the years, White ended up being the chief engineer of the Lehigh, Delaware, Raritan, and Union Canals. Later, he was instrumental in locating and engineering water supply sources for metropolitan New York. The final years of his short life were spent as president of the Cohoes Company, which specialized in the development of waterpower.

Always highly supportive of—and encouraging to—the young people under his direction, "professor" Wright was quick to praise them. At Canvass White's funeral in 1834, the gracious professor said, "It is proper that I should render a just tribute to a gentleman who now stands high in his profession, and whose skill and sound judgment, as a civil engineer, is not surpassed, if equalized, by any other in the U.S. To Canvass, I could always apply counsel and advice in any great or difficult case" (Condon 1974).

Another top assistant central to Wright's Erie operations who had a great career both as an engineer and trainer was John Jervis (1795–1885). His key projects included building the Delaware and Hudson Canals. Working under him on the Cochituate Aqueduct in Boston was Ellis Chesbrough (1813–1886), who would become one of America's first notable environmental engineers. Chesbrough was the mastermind for Chicago's historic water supply and sewage treatment systems in the late 1860s.

After Erie, Wright himself went on to engineer and/or consult on numerous canal projects in a number of states. He also became engaged in the design and construction of many of the country's early railroads and was chief engineer for the Chesapeake and Ohio Canal (1828–1831), the St. Lawrence Ship Canal (1832), the New York and Erie Railroad (1833), and the Tioga and Chemung Railroad (1836).

Wright's main assistant on the Chesapeake and Ohio project was a brilliant young engineer, Charles Ellet (1810–1862), who went on to design numerous canals for Schuylkill Navigation. He was also the first American to design and build a wire cable suspension bridge in the United States—the Fairmont, Pennsylvania Bridge over the Schuylkill River in 1842. Ellet's memorable bridge-building career earned him the nickname "American Brunel," after the great British father-son railroad, tunnel, and bridge engineering team, Marc and Isambard Brunel.

In capping off his highly productive and inspiring career, an aging Benjamin Wright successfully served as the chief engineer and street commissioner for New York City in the mid-to-late 1830s. Although he retired when he turned 70, Wright found time to chair the first engineering committee to look into establishing a national society for American civil engineers. (The committee's work created a great deal of interest nationwide and promoted discussions that led to the founding of ASCE in 1854.)

Wright's "Erie School of Engineers" and the Erie Canal, which paid for itself within two years and brought prosperity to every town along its path, raised the standards of canal building nationally and internationally. Many of the methods and tools Wright's team used were devised as the work progressed. And although those engineers doing the canal's design and construction were amateurs at the beginning, they went on to become the foremost canal builders of their day. Recognized as the country's finest civil and hydraulic engineers, they were held in high esteem and treated as national heroes.

As a tribute to Wright and his team, it was said by the media of his day: "They built the longest canal in the world, in the least time, with the least experience, for the least money, and the greatest public benefit." One tribute that recognized the judge and his accomplishments has lasted the course of time. Wright Hall at the University of Connecticut was named in his honor.

Wright died at age 72 in New York City on August 24, 1842, 6 years after the death of his devoted wife of nearly 40 years. His nonstop career spanned the reigns of America's first eight presidents—George Washington, John Adams, Thomas Jefferson, James Madison, James Monroe, John Q. Adams, Andrew Jackson, and Martin Van Buren. The legendary cadre of American civil engineers he educated, trained, and inspired—and who in turn trained and inspired their own followers—moved the nascent U.S. civil engineering community forward and upward in quantum leaps.

Stephen Harriman Long

A civil engineer so celebrated that a mountain is named after him—14,256-foot-high Long's Peak in Colorado (the tallest mountain in Rocky Mountain National Park)—Stephen H. Long was a man of rare and diverse talent. Over his illustrious 50-year career, he built railroads, bridges, forts, and dams and became expert at improving floodplains. A talented boat designer responsible for improving steamboat engines and advancing river navigation, he was the first U.S. Army *engineering* officer to take scientists deep into untouched regions of the American West in the early 1800s.

As one of nineteenth-century America's most prolific explorers, Long traveled more than 26,000 miles by land in five discovery expeditions. He explored, studied, and mapped mountains, plains, rivers, and valleys in areas previously unknown to Americans. A thorough documenter and prolific author, he wrote numerous widely read books and articles throughout his life. Long's feats as an explorer and pathfinder into the rugged American frontiers of his day echo the exploits of today's space astronauts, deemed national heroes for boldly exploring the unknown and advancing technology and scientific thought. The acclaim for his 1820 Platte River expedition, for instance, was as notable as that for engineer Neil Armstrong's stepping on the moon in 1969. The public's fascination with engineer Long's deeds was just as celebrated—and as important to progress—as the modern-day space exploits.

Born on December 30, 1784, in Hopkinton, New Hampshire, Stephen was 1 of 13 children born to Moses Long and Lucy Harriman. He graduated from Dartmouth College, receiving a bachelor's degree in 1809 and a master's degree in 1812. After a few years teaching school, he joined the U.S. Army in December 1814 as an engineer with the rank of second lieutenant.

For two years, he was a professor of mathematics at the U.S. Military Academy at West Point. In 1816, he was advanced to major in the Corps of Topographical Engineers, and, one year later, he received his first assignment to explore the boundaries of America—an 1817 expedition to the upper Mississippi region to survey the portages of the Fox and Wisconsin Rivers. One of Long's other tasks was to construct Ft. Smith in Arkansas. (Later, Long's 1860 book, *Six-Oared Skiff to the Falls of St. Anthony,* describing the Mississippi expedition became a major hit nationwide.)

Stephen Long

Photo credit: Colorado Historical Society

On March 3, 1819, Stephen married Martha Hodgkiss of Philadelphia, Pennsylvania. He was 34 and she was 20. She was the younger of two daughters of Michael and Sarah (DeWees) Hodgkiss. Stephen and Martha would have seven children, William, Henry, Richard, Benjamin, Mary, Edwin, and Lucy, the oldest being 14 years older than the youngest.

The young couple established their residency in Philadelphia right after their marriage, but duty quickly called. More than that, the lure of the western frontier was too great for the 34-year-old army major to stay put. In July of 1819, he joined General Henry Atkinson's "Yellowstone expedition," bound from St. Louis to the Rocky Mountains on the steamboat *Western Engineer*, the first steamboat to travel up the Missouri into Louisiana Purchase territory.

After building Fort Atkinson at Council Bluffs, Iowa, Long returned east to be with his new bride for the winter. In the spring of 1820, he was appointed by President James Madison to head up the first scientific reconnaissance to explore the territory between the Mississippi River and the Rocky Mountains. He and his group were given the task to find the headwaters of the Platte River and map other significant waterways flowing eastward out of the Rockies, and then return east tracing the nation's southern border along the northern edge of the Spanish colonies in North America. Because the United States had recently signed a treaty with Spain, mapping the border between the two countries was vital for America's westward movement to the Pacific.

To search for the source of the Platte River, Long embarked from the Missouri River on June 6, 1820, with 19 men, including Titian Peale, a noted naturalist; Thomas Say, a zoologist; and Edwin James, a physician who was highly knowledgeable in both geology and botany. Their scientific discoveries became valuable extra bonuses for the expedition. While finding the source of the Platte, the group reached Long's Peak, which was at the time thought to be the highest mountain in Colorado.

After traversing the Rockies southward from Long's Peak to Pike's Peak, Long's party journeyed southeasterly into Oklahoma, where groups of Plains Indians were encountered, some friendly, some not. It was the first trek along the eastern slope of the Rockies and across the Panhandle—and the first scientific survey of the area made by white men. And it gave Long the opportunity to show Native Americans the U.S. flag and tell them that the "Great White Father," President James Monroe, was a benevolent leader who would look after them.

Long's message to the central plains Indians was not dissimilar to Lewis and Clark's to Native Americans north of the 38th parallel a dozen-plus years earlier, on their "Corps of Discovery" trek from St. Louis to the Pacific Ocean. Then, though, the "Great White Father" presented was President Thomas Jefferson. Both exploratory groups found most of the Indian tribes desirous to be helpful, but not especially anxious to have the pale-faced strangers linger.

Three key results of Long's Platte River expedition were (1) an updated portrayal of Plains Indians customs and lifestyles, (2) an accurate description of the land west of the Missouri River, and (3) firm establishment of the southern boundary of the United States.

> "*The Great American Desert, the central plains from Nebraska to the eastern slope of the Rockies, is totally unfit for cultivation and unsuitable for a people depending upon agriculture.*"
>
> STEPHEN LONG

Although Long's maps of the region proved invaluable for the county's westward development, his verbal description of the land from Nebraska to the eastern slope of the Rockies made prospective settlers think twice. He called the region "the Great American Desert, totally unfit for cultivation and unsuitable for a people depending upon agriculture." He opined that the eastern wooded portion of the country should be filled up before the Republic attempted any further extension westward. He said, "Sending settlers to the area was out of the question" (Stone 1956).

From 1820 to 1823, Long finalized his reports and maps from the Platte River venture. In 1823, he led another expedition to explore the sources of the Minnesota and Red Rivers along the United States–Canadian boundary west of the Great Lakes. A few years later, he constructed the first dam (in today's terms, a wing dike) for the federal government on the Ohio River.

Brevetted a lieutenant colonel in 1826, one year later he was assigned by the War Department as consulting engineer to the Baltimore and Ohio (B&O) Railroad. In that position, Long laid out several routes and engineered countless bridges, many of them wooden covered bridges. He also formulated tables for determining railroad curves and grades, which he included in his 1829 *Rail Road Manual*—the bible of its time for the design of railroads.

Long was the first American engineer to use mathematical calculations to analyze trusses. In 1830, he invented the "X" or "Long truss," obtaining a U.S. patent for it the same year. He later patented his designs for bracing and counterbracing wooden bridges.

Long's Peak in Colorado

The majestic 14,256-foot mountain, the tallest peak in Rocky Mountain National Park that crosses the Continental Divide, is named in the honor of one of the country's most illustrious nineteenth-century civil engineers—trailblazer Colonel Stephen Long.

Photo credit: George Whitworth, Jr.

Along with his U.S. Army duties in the 1830s, Long was also a consultant to railroads. In 1834, he was named the chief engineer for the Atlantic and Great Western Railroad and surveyed railroad routes in Georgia and Tennessee.

From 1842 to 1856, he headed up the Army Corps Office of Western River Improvements. After the enactment of the Rivers and Harbors Act of 1852, he developed an intensive program for improving the waterways under his authority. In 1856, when the Topographical Engineers became a separate corps, Long was put in charge of navigation improvements on the Mississippi. In 1858, he moved his home and headquarters to Alton, Illinois, where four of his brothers had settled.

In 1861, after being promoted to colonel, he was called to Washington, D.C. and named Commander of the U.S. Army Topographical Engineers. He held the position until his retirement in June 1863, three months after the Topographical Corps merged with the U.S. Army Corps of Engineers.

The colonel moved back to Illinois after retiring. He died in his home in Alton on September 4, 1864.

Although Long lived many places during his life, he returned from time to time to his roots in Hopkinton, helping to improve bridges and other public works in the area. In C. C. Lord's *The History of Hopkinton*, it was reported that Long "was the principal mover behind local attempts at silk manufacturing." And, more importantly to some, he was responsible for the successful draining of the village frog pond, which greatly improved sanitation and land values (Lord 1890).

Theodore Dehone Judah

In the summer of 1852, while plans were being finalized for the founding of ASCE, a visionary young East Coast civil engineer was busy working out grand plans for building the country's (and the world's) first transcontinental railroad. The project would be dubbed the greatest civil engineering achievement of the middle part of the nineteenth century—the catalyst for the final push to connect the East and West Coasts of America.

Many people had talked about a railroad line to link the country from sea to sea, but the person who worked out the details showing how it was possible—the one most responsible for the mammoth project's successful completion—was Connecticut-born Theodore Judah, a daring young civil engineer with a strong entrepreneurial bent.

In his book *Men to Match My Mountains*, Irving Stone wrote, "The decade of 1859–1869 was forcibly pried open by that *rara avis*, an authentic genius, Theodore D. Judah" (Stone 1956). Civil War hero General William T. Sherman, a vice president of the Sacramento-Folsom Railroad that Judah also designed, said, "The transcontinental railroad advanced our country one hundred years."

Born in Bridgeport in March 1826, Theodore Judah was the son of an Episcopal clergyman. Young Judah graduated as a civil engineer from Rensselaer Polytechnic Institute in Troy, New York, at a time when America's booming railroads were in dire need of talented and industrious young civil engineers like Judah. When 21-year-old Theodore married pretty Anna Ferona Pierce, daughter of a successful Greenfield, Massachusetts, merchant, the trim, well-dressed, and self-confident young New Englander had his pick of jobs.

Theodore Judah

Photo credit: Institute Archives and Special Collections, Rensselaer Polytechnic Institute, Troy, New York

By the time Judah graduated from engineering college, America was well into a national "railroad fever." In 1834, the young nation had fewer than 800 miles of railroad. Within 10 years, that amount had increased to 4,300 miles. By 1854, it was 15,700 miles. And by 1864, the country's railroad trackage had increased to nearly 34,000 miles, with another 15,000 miles under construction. The United States could boast that it had more railroad miles than any other nation around the world. By May 1869, when the transcontinental railroad was completed, four of every ten railroads in the world were located in the United States. Initially Judah worked on challenging canal projects, including a section of the Erie Canal, but he quickly became enthralled with the excitement and glamour of engineering railroads. His early rail-building assignments included several cutting-edge projects: the Troy and Schenectady Railroad; the New Haven, Hartford and Springfield Railroad; the Connecticut River Railway; and Niagara Gorge Railroad —one of the most notable engineering triumphs of the 1840s.

In 1854, when 28-year-old Judah was engineering on the New York and Erie Railroad, he received an urgent telegram from New York Governor Horatio Seymour. He summoned Judah to a conference with Charles Wilson, president of a group of Californians who wanted to build a railroad from Sacramento to the rich Sierra Nevada mines. Seymour had recommended Judah as the most competent young railroad builder in the east.

Initially, Judah rejected Wilson's offer out of hand. Judah's brother Charles, who had been enticed to go west during the 1849 California gold rush, was practicing law in San Francisco. He had often written Theodore and Anna about the harsh, uninviting ways of life in the far west.

> *"People are disposed to look with distrust upon grand speculations. Convincing facts and details are needed to back them up."*
>
> THEODORE JUDAH

But after three days of negotiating, Wilson changed Judah's mind. He appealed to Judah's empire-building dreams and pointed out that a transcontinental railroad would never be possible without first solving the problem of crossing the treacherous Sierras. Once Judah finished building the Sacramento Railroad into the Sierra Nevada foothills, Wilson reasoned, it would be only logical that Judah, and Judah alone, would have the expertise to cross the mountains themselves. Within a month of that discussion, Theodore and Anna boarded a steamer heading to California and the Wild West.

Described by Oscar Lewis, author of *The Big Four*, as "studious, industrious, resourceful, opinionated, humorless and extraordinarily competent," Judah was never a man to waste motion (Lewis 1938). He completed the Sacramento project in record time and then turned his attention to a much bigger goal—finding a practical route through the Sierras to make possible the building of a railroad across the continent.

Judah quickly became the nation's most outspoken and unrelenting advocate for this seemingly impossible project. No leader in America, including the president of the United States, was as passionate—or consumed—with the notion as 32-year-old Theodore D. Judah. He constantly talked about his dream to anyone who would listen. This obsession led to the first Pacific Railroad Convention in 1859. Its main purpose was to push for federal support and funding to initiate plans for a transcontinental railway.

So impressed with Judah—and his 1859 report "A Plan for Building the Pacific Railroad," which detailed practical construction solutions—were the conventioneers that he was unanimously selected as the delegate to travel to Washington, D.C., to get Congress's approval for the project. In his report, Judah wrote, "People are disposed to look with distrust upon grand speculations. Convincing facts and details are needed to back them up." So he showed them practical solutions and facts to alleviate any negative perceptions about his bold railroad plans. Illustrative of his genius were the routes he laid out through the Sierra Nevada. They often followed ridgelines instead of the usual valleys, which would have been impractical.

Judah's lobbying efforts were favorably received by both Congress and President James Buchanan, but they wanted more facts and figures—and better cost projections. Plus, they had to consider that southern leaders threatened to secede from the Union if Abraham Lincoln were elected president in the upcoming 1860 elections. With civil war imminent, Judah's grand railroad idea took a back seat to these concerns.

Disappointed but not discouraged, Judah returned to California and busied himself developing more exacting quantity and cost estimates, maps and construction drawings, roadway slopes and profiles, and tunnel and bridge designs.

During 1860, he drew up articles of incorporation for the Central Pacific Railroad. Using his own funds and those of several close friends, Judah raised the seed money to start the company. But much more money was needed, so large blocks of stock were sold to local Sacramento merchants, mostly to four men—Collis Huntington, Mark Hopkins, Leland Stanford, and Charles Crocker. They would become famous as the "Big Four" not only of the Central Pacific Railroad, but also of California politics and commerce. Stanford would later be elected governor of California and would found Stanford University.

Golden Spike National Historic Site, Promontory Summit, Utah

When the U.S. Transcontinental Railroad linked on May 10, 1869, the news was heralded around the world; the most monumental engineering feat of the nineteenth century was completed! Shown are replicas of the locomotives involved in the first head-to-head meeting of the east and west—Central Pacific's wood-burning "Juniper" (left) and the Union Pacific's coal-burning "Engine No. 119."

Photo credit: Evelyn Weingardt

Once involved, the Big Four took control of "Judah's railroad," naming themselves its officers. Judah, though, wisely was kept as its chief engineer. And, once again, he was dispatched to Washington to lobby Congress and President Lincoln, which he did quite effectively. The Pacific Railroad Act was signed into law on July 1, 1862, and the transcontinental railroad was on its way to becoming a reality.

With his work in D.C. complete, Judah hurried back to California to continue developing the railroad of his dreams, but the Big Four had other plans. Before construction began, bitter disagreements developed—and major arguments ensued—among the five. The disagreements were primarily over the quality of the proposed road, something that challenged Judah's integrity as an engineer. The Big Four gave Judah the ultimatum either to do things their way or buy them out. He vowed to do just that and contacted financiers on the East Coast.

He left California in October 1863 for New York to raise the funds, his first meeting scheduled with Cornelius Vanderbilt. But, traveling by ship with his wife along, Judah contracted yellow fever while crossing the Isthmus of Panama. He was immediately taken to a hospital in New York. Within a few weeks (on November 2), this visionary 37-year-old engineer died, never to fully enjoy the fruits of his labor. His enthusiasm and advocacy for the transcontinental railroad made it a reality long before its time. It was a staggering civil engineering accomplishment that forever transformed the U.S. economy and spearheaded the Industrial Age in America, yet he never saw a single rail laid for it. When the final connection of the transcontinental railroad was made on May 10, 1869, at Promontory Summit, Utah, many took their bows during the celebrations as the final "golden" spike was driven. Unfortunately, Judah was not one of them.

Two other skilled civil engineers, however, were immortalized for posterity that day. Shown shaking hands in "center court" of the famous Andrew J. Russell photograph of the event were Grenville Dodge with the Union Pacific and Samuel Montague with the Central Pacific, Judah's replacement.

William Le Baron Jenney

So influential was William Le Baron Jenney in training Chicago's emerging master architects in the late 1800s that he is most often remembered as an architect rather than the outstanding civil engineer he was. Unmistakably, though, the Civil War veteran continues to be credited as being the father of the modern skyscraper after designing Chicago's Home Insurance Building, which was completed in 1885.

In its day, it was the world's tallest structure using a metal frame rather than the traditional masonry-bearing walls. The frame, originally designed in iron, was changed to steel midway through construction when Jenney found that the Carnegie-Phipps Steel Company could produce high-quality steel beams and columns. It was the world's first commercial building to make use of structural steel in its frame. While the 180-foot-tall structure was being constructed, a confident, 52-year-old Jenney declared, "We are building to a height to rival the Tower of Babel." This project provoked a never-ending high-rise construction rage around the world, each new building aspiring to be higher than its neighbor.

With his star pupils, architects Louis Sullivan, Daniel Burnham, William Holabird, and Martin Roche (all attained international stature in the field of architecture; Sullivan became Frank Lloyd Wright's mentor), Jenney formed the nucleus of the world-renowned "Chicago School" of architecture. "His bold vision and utilitarian aesthetic became the cornerstone of

Chicago's commercial architecture," reported Judith Dupre in *Skyscrapers* (Dupre 1996).

Jenney was born on September 25, 1832, in Fairhaven, Massachusetts. His father was a prosperous New England shipowner. Being raised among such enterprise and wealth contributed to William's adventuresome nature; he sailed around Cape Horn, visited Manila, and enrolled in one of Europe's most elite universities before he was 20 years old. Because of his background and temperament, it came as no surprise to peers and contemporaries that Jenney developed into one of the world's foremost pioneers of building technologies before he was 40 years old.

William Jenney

Photo credit: Chicago Historical Society

Jenney first studied civil engineering at the Lawrence Scientific School of Harvard University and then entered the renowned Ecole Centrale des Arts in Paris in 1853. He graduated three years later in 1856—one year after his classmate Gustave Eiffel (of Eiffel Tower fame). He decided to move back to America immediately after graduation from the Ecole. However, he returned to Paris two years later for more training following work on a railroad project in Panama. After fewer than 18 months in the City of Light—with the outbreak of the U.S. Civil War imminent—William rushed back to his native country to join the Union Army.

In 1861, Jenney came under the command of General William T. Sherman as a civil engineer with the rank of major. As Sherman's chief engineer, he was responsible for designing numerous engineering works, including fortifications for strategic battles such as those at Corinth, Shiloh, and Vicksburg. The general also used Jenney's engineering talents to direct demolition work in conquered regions. Participating in Sherman's famous "March to the Sea," Jenney learned how to destroy bridges and buildings with great efficiency. The experience taught him much about iron construction, foundation, and framing techniques—know-how that would prove to be invaluable later in his career.

After the war, William returned East to work as a coal-mining engineer, but by then his true ambitions were to build structures rising above the ground. In 1867, he migrated to bustling Chicago, where he founded his own civil engineering and architectural design firm. He based the firm on dividing tasks among specialists, which enabled it to undertake a then-unheard-of number of large projects all at once. The "Major," as Jenney was called, kept his young designers like Sullivan and Roche busy working on façades while he himself concentrated on civil and structural engineering problems. Said Sullivan, "Jenney was an architect only by courtesy of terms. His true profession was that of an engineer" (Duis 1976).

A frequent contributor to the periodical *Inland Architect*, Jenney became an extremely effective spokesman for the industry, helping found the Western Society of Architects in 1884. According to Perry Duis, the author of *Chicago: Creating New Traditions*, "Jenney warned young archi-

*The world's first skyscraper—
the Home Insurance Building,
Chicago, 1885*

The structure commenced the era
of modern skyscrapers, opening the
door to "the sky's the limit" thinking. It was the first to have an iron
and structural steel frame as its main
structural system, overcoming the
restrictions masonry-bearing wall
construction placed on buildings.

Photo credit: Chicago Historical Society

tects [and engineers] that they
should not bend too easily to
the whims of the client, even
though they paid the bill. Part
of an architect's [and engineer's] job was to try and up-
hold the canons of good taste.
Yet, at the same time, the architect [and engineer] had to
fit a building to the owner's
genuine needs."

Jenney was also a real
stickler for public safety and
clarity of design, writing, "No
arrangement [building] for the
health, convenience and pleasure of the occupants should be
sacrificed to any ideas of sym-
metry or external effect." He
believed that ornament should
not be added to buildings un-
less it had a definite purpose
or clearly reflected the activity
inside—and materials should
never be camouflaged to look
like something they were not.
Sullivan learned these lessons
well. He passed them on to
Wright, who became legendary for this concept.

Major Jenney's talents were immensely broad; he
was a pioneering civil and structural engineer, architect,
landscape and urban designer, town planner, teacher,
and city leader. A prolific writer, he did not limit his
writings to engineering, architectural, and construction
issues. He was the author of popular papers on a wide
range of subjects, including "The Fossils of History,"
"Personal Reminiscences of Vicksburg," and "The Age
of Steel."

In addition to Home Insurance, key Chicago buildings designed by Jenney included (1) Manhattan Building (1891), the first building to achieve a height of 16
stories; (2) Leiter (Sears and Roebuck) Department Store (1892), which was
for several years the largest department store in America; (3) Horticultural
Building (1893), the largest botanical conservatory ever built; and (4) the
ultramodern, steel-framed New York Life Insurance Building (1898).

On May 8, 1867, Jenney and Elizabeth "Lizzie" Hannah Cobb, from
Cleveland, Ohio, were married. He was 34 and she was 20. Her father

was a bookseller who admired classical culture so much he named Lizzie's brothers after Roman heroes. The Jenneys would have two sons, Max and Francis, neither of whom would follow in their father's footsteps and become engineers. It has been reported in early stories written about him that Jenney said he got his inspiration for framed construction from his wife's birdcage, watching her move it from place to place in the house while she was cleaning (Turak 1986).

The major died in Los Angeles on June 15, 1907, at age 74. His ashes were scattered over the grave of his devoted wife, who preceded him in death by nine years and was buried in the famous Graceland Cemetery in Chicago.

William Jenney left a colossal legacy, not only as the creator of the skyscraper and the founder of the Chicago School of Architecture, but also as a world-class town planner and civil engineer. For example, with Frederick Law Olmsted (the designer of New York's Central Park), Jenney helped create the plan for Riverside, Illinois, the nation's first planned "railroad suburb." Included in the venture was an ornate water tower and rail depot befitting the trendy development—one of the nation's first resort and residential mixed-use developments. In Chicago, he also designed the West Parks section of town, where broad tree-lined boulevards link an extensive system of connecting lake-filled parks. Most notable among them are Douglas, Garfield, and Humboldt Parks.

> "*Ornament should not be added to buildings unless it has a definite purpose or clearly reflects the activity inside, and materials should never be camouflaged to look like something they were not.*"
>
> WILLIAM JENNEY

In the 1998 book *1,000 Years, 1,000 People: Ranking the Men and Women Who Shaped the Millennium*, Jenney is ranked number 89, well ahead of the likes of Alfred Nobel, Catherine the Great, Queen Victoria, and Henry VIII of England. According to its authors, Agnes Hooper Gottlieb et al., "Ever since [Jenney's Home Insurance Building], carving space from the sky has soared in crowded cities, and the skyscraper has become the symbol of the future" (Gottlieb 1998).

CHAPTER TWO

Environmental Experts

The driving goal for this millennium is environmental, economic and social sustainability. Pursuing our professions and businesses in ways that can be sustained without denying future generations their opportunities must become a bedrock principle.

—General Hank Hatch
(former head of the U.S. Army Corps of Engineers)

Much of what this generation and past generations have done on earth, especially with regard to the built environment, has been for the better. However, some activities, not the least of which is the world's ever-continuing population growth itself, have substantially added to the contamination and pollution of the planet's waters, air, and land, both above and below the surface.

In mitigating such issues and preventing future problems, the single group most needed is civil engineers. They are, in essence, the protectors of earth's environment and everyone's standard of life. They are also indispensable in ensuring that today's designs and construction projects incorporate sustainable development criteria in a pragmatic way. Future generations and civilizations will be greatly affected by how successfully today's and tomorrow's civil engineers resolve these issues. To be as effective as possible in the future, though, requires that increasingly more civil engineers take on top leadership positions in groups controlling city and state planning and municipal and environmental engineering.

This chapter looks at the lives of four civil engineering icons—Ellen Richards, Holly Cornell, William Moore, and Fu Hua Chen—who were noteworthy trailblazers in remedying environmental issues and/or developing parameters for optimizing the use of earth's soils to support structures and to absorb the by-products of the built environment.

The foremost pioneer in the investigation, research, and development of design criteria for soils, surface and subsurface, was Karl Terzaghi (1883–1963), the "Father of Soil Mechanics." The instigator of soil mechanics as a new science, Terzaghi first introduced his findings in Vienna in 1925 while working at Istanbul Technical University and the Bogazici University in Istanbul, Turkey. During his Vienna presentation, he delineated innovative theories on soil consolidation and settlement, bearing capacity, shear strength, slope stability, lateral earth pressures, and retaining wall criteria.

He not only produced theories but also developed practical approaches to applying them, and his books and papers included hands-on design examples and design/analysis charts. He summarized his main soil theories and designs in his seminal text *Soil Mechanics Based on Soil Physics* (1925) and further refined them in *Theoretical Soil Mechanics* (1943). Both became the "gold standards" of the industry.

In 1938, after Germany took over Austria shortly before the beginning of World War II, Terzaghi left his homeland for America and Harvard University. Over his career, his consulting work and activities took him all over the world, and he mentored and inspired countless followers. He frequently collaborated with some, like Rensselaer Polytechnic Institute's soil guru Ralph Peck (born 1912). They produced publications that further refined and updated industry standards and the latest applications of soil mechanics. Their textbook *Soils Mechanics in Engineering Practice* became the classic for the industry. Even after his retirement in 1956, Terzaghi continued to share his knowledge by lecturing around the globe until his death, which was shortly after his eightieth birthday.

In addition to Terzaghi and Peck, two other soils/geotechnical engineering greats clearly stand out—Moore and Chen. They are highlighted here, not only for their pioneering geotechnical, hazardous waste and contamination, and mining and land reclamation innovations, but also for the tremendous impact they had on the profession worldwide. Their colorful and inspiring lives touched people all around the world.

The two other engineering giants profiled—Richards and Cornell—were also singled out because of their unwavering courage and history-altering contributions while pioneering changes within the profession. Richards was a forerunner in the teaching of environmental, sanitary, hydraulic, and water resources engineering, while Cornell was instrumental in the industry for establishing and refining it. Both raised the respect for these activities to the highest rung within the field of civil engineering and in the public's eyes.

Ellen Swallow Richards

In 1870, when Ellen Henrietta Swallow became the first woman admitted to the Massachusetts Institute of Technology (MIT), eyebrows were raised. When she received a B.S. in chemistry three years later, precedents were set. American women across the country started developing an interest in—and actually receiving—scientific degrees from U.S. institutions of higher learning.

Three years after Ellen received her chemistry degree, Elizabeth Bragg Cumming (1859–1929) became the first woman to earn an engineering degree from an American university—a B.S. in civil engineering from the University of California at Berkeley in 1876. Others followed suit, like Lillian Moller Gilbreth (1878–1972), possibly the most well-known women engineer in history, in part because of the widely distributed Hollywood movie "Cheaper by the Dozen." Lillian earned her Ph.D. in psychology and became an expert in industrial engineering.

Ellen Swallow Richards

Photo credit: Sophia Smith Collection, Smith College

Over her career, Ellen Swallow Richards made numerous contributions toward improving U.S. sanitary conditions and standards of living by applying her scientific knowledge of the properties of chemicals and materials and by teaching sanitary engineering to America's first wave of environmental engineers. Her innovative analysis of water and mineral samples set industry standards, and her practical application of the applied sciences became the driving force behind the home economics movement in the United States at the turn of the century. Because her work greatly improved household and industry sanitation—and the public's environmental awareness—one of her biographers, Robert Clarke, deemed her the "woman who founded ecology" (Shearer 1997).

Honoring her as one of the top environmental engineering leaders of the last 125 years, *Engineering News-Record* (ENR 1999) labeled Richards as "the first female environmental engineer," adding, "She refused to be intimidated by her male counterparts. She produced the world's first water purity tables in 1877, and helped establish the first systematic course in sanitary engineering—at MIT. Her work also led to the establishment of the Woods Hole Marine [Biological] Laboratory in Massachusetts."

Ellen was born on December 3, 1842, in Dunstable, Massachusetts, the only child of Peter and Fanny (Taylor) Swallow, both of whom were schoolteachers. In her early years, she was home-taught. When her family moved to Westport, Massachusetts, she attended the Westport Academy from 1859 to 1863. After her parents moved to Littleton, Massachusetts, in 1863, where, in addition to teaching, her father ran a store, Ellen began saving money in earnest to further her education. To this end, she worked at a variety of domestic jobs, helped her father in his store, tutored students, and taught elementary school.

In 1868, 26-year-old Ellen Swallow was accepted at Vassar College, graduating in 1870 with a B.A. degree in chemistry. She was then accepted at MIT, where she earned her B.S. degree in the subject. The same year, she received a master's degree in chemistry from Vassar. She then continued advanced studies at MIT and took a position as a teacher in its department of chemistry, frequently without pay.

After a two-year engagement, she married Professor Robert H. Richards, head of the MIT Department of Mining Engineering, becoming Ellen Henrietta Swallow Richards on June 4, 1875. The couple would have no children, but they frequently collaborated on a variety of provocative scientific and teaching assignments. Ellen's frequent consulting work for Robert on the chemistry of ore analysis was instrumental in her being elected to the American Institute of Mining and Metallurgical Engineers in 1879, becoming its first female member.

In 1876, Ellen convinced the Women's Education Association of Boston to provide funding to open a women's scientific laboratory at MIT, where women were given the opportunity to enter scientific fields. There they learned basic chemistry, biology, and mineralogy, which allowed a good number of them to secure high-level industrial and government consulting jobs.

In 1882, Richards cofounded the Association of Collegiate Alumnae (later known as the American Association of University Women).

> *"One of the most serious problems of civilization is clean water and clean air, not only for ourselves but for the planet."*
>
> ELLEN RICHARDS

From 1884 to 1911, she served as Professor William Nichols's top assistant—and as an instructor—in MIT's new laboratory of sanitation chemistry and engineering. She taught air, water, and sewage analysis, and introduced biology to MIT's curriculum. As one of MIT's leading professors of sanitary engineering, she educated the men who went on to design and operate the world's first modern municipal sanitation facilities, encouraging them to become "missionaries for a better world" (Kass-Simon 1990).

Richards and her colleagues, no doubt, were well aware of—and took into account while teaching environmental and sanitary engineering—the successes Chicago was having with its modern sewage treatment and water supply systems, designed by the innovative civil engineer Ellis Chesbrough (1813–1886). Completed in 1869, Chicago's sanitation and water supply innovations were saving countless lives. Deaths from cholera at the rate of 2,000 deaths per 100,000 inhabitants were the norm in the 1850s; once Chesbrough's engineered sewage and water systems were operational, cholera outbreaks in Chicago became nonexistent (Griggs 2003).

The efforts of Richards and Chesbrough set the stage for other leading environmental engineers who followed them. Both greatly impacted the work of all early twentieth-century practitioners and teachers, professors like the legendary Abel Wolman (1892–1989) from Johns Hopkins University. Over his 75-year career, Wolman taught and mentored thousands of students in the practice of sanitary engineering as it is known today. Many of his developments, like innovative water and sewage chlorination and disinfection procedures, found application globally. A true "citizen of the world"—and extremely active in his community and the public arena—Wolman wielded strong influence in federal policy on water pollution control and water resources management.

Woods Hole Marine Biological Laboratory in Massachusetts, 2004

Ellen Richards's work—including producing the world's first water purity tables in 1877 and helping establish the first course in sanitary engineering at MIT—led to the establishment of the Woods Hole facility, currently a cutting-edge, world-class environmental research operation.

Photo credit: Andrew Hawkins/Marine Biological Laboratory

Among the lasting original contributions to the environmental and chemical fields Richards developed while at MIT are the "normal chlorine map," which serves as an early warning system for inland water pollution, and the "water purity tables," the first water quality standard in the United States. In addition to her work at the MIT sanitation laboratory and investigating water pollution and designing safe sewage systems, Richards tested home furnishings and foods for toxic contaminants. She also consulted, lectured, and wrote books, papers, and bulletins on nutrition for the U.S. Department of Agriculture.

A chance meeting with industrialist Edward Atkinson, president of the Manufacturers Mutual Insurance Company, in the mid-1880s happened while Richards was doing a comprehensive survey of factories for fire prevention purposes. This led to the two successfully teaming together on a number of projects.

One venture was the design of fire-resistive factories, which were copied throughout the country. To help reduce the frequency of industrial fires, Ellen invented a noncombustible oil for machines. She ultimately became an authority on industrial and urban fires and became a sought-after expert on analyzing schools and other public buildings for fire resistance.

In 1887, Richards was hired by the State Board of Health to survey the quality of water in Massachusetts. Her work resulted in suggestions for correcting water pollution, and led to the nation's first set of state water quality standards. For 10 years thereafter, she was employed as the official water analyst for the board. She also served as dietary consultant to a number of hospitals, both locally and nationwide. This led to her involvement in the 1893 World's Columbia Exposition in Chicago, where she created the "Rumford Kitchen," a concept that provided inexpensive but nutritious meals for working-class men and women. This kitchen helped inform the public about nutrition and safe food preparation.

Her work dealing with improving the nutritional and sanitation practices for ordinary citizens reached a high point in the summer of 1899 when she organized and chaired a national conference in Lake Placid, New York. The gathering helped define nationwide standards for teacher training and certification in the new field of home economics. In 1908, key attendees at the conference formed the American Home Economics Association, electing Richards its first president.

A stickler for maintaining a clean environment inside buildings as well as outside, Richards was a pioneer in identifying and addressing what modern-day health officials call "sick building syndrome." After a comprehensive survey of MIT campus buildings, Richards informed MIT President Arthur Noyes (1907–1909) that the heating and ventilation systems in several MIT buildings were so flawed that their interior environments posed serious health hazards needing immediate remedy—and she prescribed the remedies. Richards believed this: "One of the most serious problems of civilization is maintaining clean water and clean air, not only for ourselves but for the Planet."

In 1910, Richards started the *Journal of Home Economics*. At the same time, she was named to the council of the National Education Association and was charged with overseeing the teaching of home economics in the nation's public schools.

Although Ellen never received her Ph.D. degree from MIT (some would say her "well-deserved doctorate"), she was awarded an honorary Ph.D. degree from Smith College in 1890 when she was 48.

During her career, Richards, whose lifelong hobbies included gardening and traveling, wrote more than a dozen books and numerous papers and articles on a number of subjects. Among her books were *The Chemistry of Cooking* (1882), *Home Sanitation: A Manual for Housekeepers* (1887), *Laboratory Notes on Industrial Water Analysis: A Survey Course for Engineers* (1908), *Conservation by Sanitation* (1911), and *Euthenics: The Science of Controllable Environment* (1912).

Ellen Swallow Richards, a true pioneer in the science and engineering of the environment and in opening up the fields of scientific inquiry for U.S. women, died on March 30, 1911, in Boston. She was 68. She left an amazing legacy as one of the nation's authorities on environmental and sanitary engineering, water and air purification, the analysis of food and human diet, and the design of safer and healthier buildings.

Holly A. Cornell

Holly Cornell was born in 1914 in Boise, Idaho. His father, Harvey Baeff Cornell, was a manufacturer's representative for the Kelly Clark Company, a producer of canned goods and groceries. He loved to tinker with his hands and build intricate model ships. Young Holly inherited these talents and also an aptitude for mathematics. The two often worked together making models of planes and ships and once built a crystal radio, which was considered a high-tech item in those days.

During the Great Depression, the Cornell family—Holly, his brother and sister, and their parents—relocated to Portland, Oregon, where he attended Grant High School and was president of the student body. After graduation in 1932, he landed a job as messenger for the Bank of California at a salary of $65 a month—outstanding Depression time wages. He used $20 for living expenses, sent $20 home to help out, and saved $25 as seed money for a college education.

Holly Cornell

Photo credit: CH2M-Hill

In 1934, Cornell enrolled at Oregon State College (OSC), a little older and more serious than the average freshman, intent on becoming an engineer. He wanted to build bridges because, he said, "They just seemed romantic." Once at OSC, however, the direction of his career path took a different slant, mainly because of an exceptional sanitary engineering professor named Fred Merryfield.

According to Cornell, "He was a wild one with this tremendous enthusiasm. He was kind of a tough instructor but a very outgoing, outspoken and strong one. In a sense, he entertained you. He put on a show when he taught." Years later, this enthusiasm, magnetism, and strong individualism (and Merryfield's high regard for three of his former students, Holly included) would be instrumental in founding one of the world's largest engineering firms—CH2M-Hill—in the unlikely college town of Corvallis, Oregon.

After graduating from OSC in 1938 with a bachelor's degree in civil engineering, Cornell attended Yale University on a graduate fellowship, partly arranged by Merryfield. Before leaving for the Ivy League school, Holly married his college sweetheart, Cleo Ritner, the day before Christmas in 1938. They would have two children, Cynthia and Steven.

At Yale, Cornell encountered another extraordinary engineering professor who would greatly influence him—Hardy Cross. Said Holly, "I think Hardy Cross was probably the best teacher I ever studied under. He was a philosopher as well as an engineer, and brought humanities and engineering ideas together in his teaching." It was under Cross's tutelage that Cornell's professional engineering interests crystallized and, in his words, "was the reason I developed a great love for engineering."

During his two years at Yale, he earned his master's degree while teaching engineering classes. On occasion, he and OSC classmate Jim Howland, who was attending nearby MIT, would get together and talk about the future. The two also kept in contact with Merryfield and another good friend from their OSC days—Burke Hayes.

After receiving a master's degree from Yale, Cornell went to work for Standard Oil in San Francisco. About eight months later, in 1941, at the height of World War II, Cornell, who had received a reserve commission at OSC, was called into the U.S. Army to serve in the Corps of Engineers. There he taught engineering courses and became a platoon leader, a company commander, and eventually an executive officer.

He served with distinction in the Corps in Europe, receiving the Bronze Star. As an executive officer under General George Patton, Cornell's engineering group repaired Germany's famed Remagen Bridge, enabling the Allied Forces to cross the Rhine.

> *"The companies that advance aren't afraid to fail. Have at it. Try it. If it doesn't work, throw it out. Try something else."*
>
> HOLLY CORNELL

At the war's end, Merryfield, the first of the four discharged from military service, returned to OSC and resumed teaching. When he was not in the classroom, he was providing consulting services to nearby communities. Recent Oregon legislation required communities to clean up wastewater outflow, and Merryfield quickly had more consulting work than he could handle. He and Cornell intensified their correspondence.

Soon, Howland and Hayes joined them and the four were exchanging a flurry of letters and telegrams discussing an engineering partnership. One of the letters (August 26, 1945) from Cornell to Howland and Hayes read:

> Fred is all wound up in a stream pollution survey on the Willamette River. Oregon has a new law that Fred is very enthusiastic about. It is a big start toward cleaning up Oregon streams. It means, according to Fred, that almost every town and city will have to treat its sewage. All the cities have plenty of money now and are begging for engineers so they can get started on post-war construction. Fred named a half a dozen jobs that he could get right now that looked like enough to keep us busy for a good six months. I, for one, am frankly enthusiastic about the idea and certainly want to give it [their partnership] a chance.

The clincher for the formation of the company came shortly after, on September 15, 1945. Merryfield sent his three former students a persuasive letter that read: "It is my reaction that water filter plants and sewage treatment plants will be in demand throughout the Pacific Northwest for the next ten to fifteen years. Many of the present gravity systems of water are on their last legs and many sewer systems are in pitiful condition."

First came Cornell, then Howland and Hayes in short order, and in 1946, tiny Corvallis was witness to the creation of the fledging firm of Cornell, Howland, Hayes and Merryfield (CH2M). Its first office, a small space above what today is a sporting goods store, was quickly outgrown and then some.

After more than two decades, CH2M merged with Clair Hill and Associates of Redding, California, in 1971, creating the company's current name CH2M-Hill.

Foothills Water Treatment Plant, Colorado, 1986

Metro Denver's first modern-day 500 MGD water purification facility at the base of the Rocky Mountains. It has allowed the Queen City of the Rockies to meet the potable water needs of its ever-increasing population.

Photo credit: ACEC-Colorado

Part of Cornell's guidance of the firm through its crucial stages included setting up and managing its first metropolitan office in Seattle. About the office, CH2M-Hill Chairman Emeritus James Poirot, who worked with Cornell from 1958 to 1978 and consulted with him for another 10 years, said:

Holly asked me to move from Corvallis to Seattle with him in 1960 to open a new office. The office was started in downtown Seattle because Holly said we needed to be where the financial decisions were made and major business opportunities were initiated. He joined the Rainier Club to be able to have proper business luncheons and asked me to join the Washington Athletic Club. We often had meals together at each club and always met city attorneys, bond counsels, city councilmen, municipal bond advisors and good friends at these clubs. He was a 'classy' guy and wanted his firm to be a 'classy' company.

Cornell believed this: "The companies that advance—as compared to the ones that stay in the same rut and lose their market share—are the ones that are always experimenting, always trying something new, aren't afraid to fail. Have at it. Try it. If it doesn't work, throw it out. Try something else."

CH2M-Hill's current president and CEO, Ralph Peterson, who joined the firm in 1965 and worked with Cornell until his retirement, said, "Holly brought the company into the computer age. But what I remember most about him are the lasting relationships he forged with clients; clients and projects like the Boeing 747 Assembly Plant in Everett, Washington, and the $100 million Denver Water Board's Foothills Water Treatment Plant. These became landmark projects, but what is truly impressive is that those clients are still valued CH2M-Hill clients today."

It was during Cornell's tenure as corporate president (1974–1978) that CH2M-Hill secured its largest contract to that point—a $1.6 billion, 12-

year pollution abatement project in Milwaukee, Wisconsin. After his stint as president, he served as chairman of the firm until 1980, when he retired.

In addition to his strong work ethics, said Poirot, "Holly was a strong promoter of professionalism and active with several engineering organizations. He was president of the Seattle Section of the ASCE, chairman of the ASCE Pacific NW Council and president of the Consulting Engineers Council of Washington. Helping him on various committees motivated me to follow in his footsteps." (Poirot was the 1994 national president of ASCE.)

Cornell, who loved golf, traveling, and classical music, died on July 1, 1997, at age 83. He left a legacy as founder of one of the nation's largest employee-owned companies that, at the time of his death, had more than $1 billion a year in business and more than 7,000 employees in 120-plus offices worldwide. Today this leading global environmental services firm is headquartered in Denver, Colorado.

William Wallace Moore

" I've always wondered if some of my enthusiasm for professional activities and engineering societies is inherited from my dad's love of preaching and telling people what to do!" reflected William "Bill" Moore in a 1998 interview. He was 86 at the time, reminiscing on his 60-plus years as an engineer and cofounder of the international engineering giant Dames and Moore (D&M), one of the world's leaders in geotechnical and environmental engineering and applied earth sciences. By 2002, D&M had grown from two partners to a complex, multifaceted operation with more than 100 partners and 6,000 employees in several dozen offices worldwide. (In 2003, D&M was absorbed by the giant URS Corporation.)

The older of the two sons of Leon Wallace Moore and Nellie Munson from the farmlands of Iowa, Bill Moore rose to the highest heights in his chosen field, first on the West Coast and then nationally and internationally. He was the first U.S. engineer to be elected president of the International Federation of Consulting Engineers (FIDIC), headquartered in Geneva, Switzerland. His brilliant performance as the 1970–1972 president of the powerful 50-plus-nation organization greatly elevated the stature of U.S. companies internationally.

Moore's grandfather, Wallace Moore, returned home from the Civil War, minus an arm, to become postmaster—and a farmer—in rural Iowa, where Bill's father, Leon, was born. Although an engineering graduate from Cornell College in Iowa, Leon's heart was not completely in it as his life's work. After a short stint working as a railroad engineer in Mexico following the Spanish-American War (1898–1899), he enrolled in a Methodist seminary and became an ordained minister.

Because the pay for ministers in the early 1900s was not high, Leon and Nellie were always hard-pressed for money, and they moved around a lot. So Bill, who was born in Pasadena, California, in 1912 and his younger brother, Walter, were always on the go as kids, attending a variety of schools.

To make more money to help his two sons through college, Leon put his ministry on hold in 1928 and took a job in the Los Angeles County engineer's office. Both boys augmented their parents' financial assistance with odd jobs (including picking pears and peddling oranges door-to-door) to earn their engineering degrees. Bill earned B.S. and M.S. degrees in civil engineering from the California Institute of Technology in 1933 and 1934.

William Moore

Photo credit: William Moore, Jr.

He immediately went to work for the U.S. Coast and Geodetic Survey measuring ground movements for its earthquake research program. Following that, he worked for the U.S. Corps of Engineers and a private engineering firm, Labarre and Converse—foundation-engineering consultants.

In 1938, Bill and Trent Dames (a Caltech classmate) boldly threw caution to the wind and founded D&M, taking on a wide array of assignments as consultants in the newly emerging field of geotechnical engineering. Many of the firm's first assignments developed from the pair's close ties with West Coast structural engineers who were advancing state-of-the-art seismic design, prompted by a major Long Beach earthquake in the 1930s.

By 1969, D&M had become well established and respected, with numerous impressive projects under its belt. That was the year the firm received a lucrative contract for site exploration and geologic studies, followed by the development of seismic design criteria for the massive Trans-Alaska Pipeline project.

In 1968, large crude oil reserves had been discovered on Alaska's North Slope (Prudhoe Bay). Within a year, plans were underway to build a pipeline south from the bay over 800 miles of frozen tundra, boreal forest, raging rivers, and majestic mountains to a marine terminal in Valdez (the northernmost ice-free port in North America). From there, tankers would transfer oil to mainland U.S. refineries.

Actual construction on the Trans-Alaska Pipeline, which began April 29, 1974, took three years and two months to complete. It opened for business on June 20, 1977, at a cost of $8 billion. At times, more than 28,000 workers furiously toiled around the clock in extreme weather conditions to finish it. The 48-inch diameter pipeline—one of the largest and first environmentally sensitive major construction projects in the world—is like no other artificial wonder. It is elevated above ground in some locations and buried in others. It is built to withstand permafrost, earthquakes, and 100-mile-per-hour winds and is capable of withstanding enormous expansion and contraction movements due to extreme temperature changes.

Upon the successful completion of the Trans-Alaska Pipeline assignment, D&M's portfolio of international projects rapidly grew, in both number and significance.

Over the years, Bill became widely known within and beyond the profession as a pacesetter in geotechnical and environmental engineering and earthquake foundation analysis. His pioneering work to improve soils in reclaimed tidelands allowed major structures and industrial complexes to be founded economically on reclaimed land without the use of expensive pilings.

Recapping his career in interviews in the late 1990s, Moore said he viewed engineering as a people business: "I've always enjoyed the technical work, but the brightest place in my recollection are the people I've met and known. We've [he and his wife] gotten to know so many people in this business and made so many lasting friendships, that has to be the highlight of my career."

According to Moore, the two most significant things impacting the profession today are (1) the "disruptive liability atmosphere" that engineers must work under, and (2) the development of the "environmental syndrome, which has required a lot of additional engineering studies and supplementary applied scientific studies." Today's engineers are working in a fishbowl, making them open to a lot of outside criticism, some well-meaning, some not.

"In this country, there is a propensity to go to court at the drop of a hat," said Moore.

> *"What we have to do is develop sound, technically trained people who are articulate enough, and patient enough, to explain technical issues to non-engineers."*
>
> WILLIAM MOORE

One cannot practice engineering without considering the possibility of a lawsuit. Engineers have to be defensive. And they can't afford to be 'just an engineer' any longer. It's not enough to figure how to design a project and make it work. Now, a project must be sold politically to a wide cross-section of nontechnical people.

Because of that, engineers can't live apart from society; they must be able to communicate with the public and influence public decisions. Therefore, we have to develop sound, technically trained people who are articulate enough and patient enough to explain technical issues to non-engineers.

Engineers by their nature are not publicly orientated. They are not given to popular persuasion—but the successful ones of the future will be the ones who develop that ability. The engineer who decides to stay out of public discussion on issues impacting his or her profession will be reduced to a non-participant in the key decisions. And that is not good for anyone—neither engineers nor society.

Moore also felt strongly about young engineers who know little or nothing about the history of the profession and do not care to understand what went before them. He believed they'd be better engineers—and a lot prouder of the profession and themselves for being in it—if they knew more about what previous generations of engineers accomplished.

His son Bill, Jr.—for many years, the manager of D&M's Cairo, Egypt, office—said his father always stressed the need to take care of business first. Bill, Sr. would say, "The way you develop a business is by using shoe leather." He often admonished, "Never eat lunch alone, because if you do, pretty soon you don't eat lunch." To him, lunchtime should be spent with a prospective client. "Spend the day with clients and do your work at night," he advocated.

Trans-Alaska Pipeline showing the crossing over the Gulkana River

Completed in 1977, the 800-mile-long pipeline extending from Prudhoe Bay to the Port of Valdez fostered state-of-the-art environmental mitigation engineering techniques. The project was prompted by the 1967 discovery of the largest oil and gas reservoir in North America, a supply capable of meeting 20 percent of the nation's oil production.

Photo credit: Peratrovich, Nottingham and Drage, Inc.

Throughout his professional life, Moore was honored with numerous awards and served on many professional and community boards and advisory committees, including the National Academy of Engineering. In addition to FIDIC, he was elected president of several other prominent engineering groups: the American Council of Engineering Companies (1964), the San Francisco section of the American Society of Civil Engineers (1958), and the Structural Engineers Association of California (1947).

Moore was the first recipient of the prestigious Arthur M. Steinmetz Award (given by *Consulting Engineer* magazine in 1981) "for a distinguished career in consulting engineering." He set the standard for all the award recipients that followed.

Although an incessant world traveler for years, his favorite nonengineering passion, right until the end, was sailboating on San Francisco Bay. This 90-year-old engineering great passed away peacefully at home in San Rafael, California, on October 23, 2002. He was survived by his adored wife of 69 years, Genie, and their three children, Bill, Jr., Roy, and Susan, and their families.

Fu Hua Chen

Honored early in his career as the man behind the incredible completion of China's Burma Road during World War II, Chen became known worldwide as the foremost authority on expansive soils.*

---

*Expansive soils are those that expand in volume—and create outward pressures—when saturated. These forces, mostly unwanted, cause structural elements like footings and floor slabs bearing on the soils to heave and move upward and/or sideways, which is often highly destructive and detrimental to the integrity of a structure.

Born in Fu Zhou, Fu Jain Province, China, in 1912, he spent his first years in Shanghai and Beijing during the turbulent times following the Guomindang overthrow of the Manchu Dynasty.

In 1932, Chen left his homeland and enrolled at the University of Michigan. After receiving his bachelor's degree in civil engineering, he went on to the University of Illinois for his master's degree. He then spent a year in Europe, where he and Karl Terzaghi became friends before he returned to China in the late 1930s.

After several years with the Chinese Highway Administration, Chen was assigned to the 720-mile-long Burma Road Project, eventually becoming its chief engineer. This was during World War II and Japanese soldiers constantly harassed and attacked workers along the road, a highway many consider to be "China's lifeline." For his leadership in successfully completing the road, one of the greatest engineering feats of all time, Chen received a prestigious national award from General Chiang Kai-shek, China's leader until he was ousted by the Communists in 1949.

Prior to his assignment to the Burma Road Project, Fu Hua met and married Edna Yu in 1941. His colleagues said, "It was a match made in heaven." And one that lasted until Fu Hua's death in 1999. All three *Fu Hua Chen* of their children were born during war-torn times— Photo credit: ACEC-Colorado Dorothy and Tyrone during World War II and their youngest, Yvonne, shortly after the atomic bombing of Japan that led to its surrender.

Just after the end of World War II, as civil war tore China apart and the Communist succession was imminent, the Chens hurriedly embarked for Hong Kong, arriving with no more than they could carry. There, Fu Hua found employment with the Public Works Department, where he stayed until 1957. That's when he left Hong Kong—and China—for good, immigrating to the United States for a job with Woodward-Clyde-Sherard (WCS), geotechnical engineers, in Denver, Colorado.

In 1961, Chen left WCS to form Chen and Associates, a private consulting firm. During the next 27 years, he developed his Denver-based company from a one-man, garage operation into a nationally recognized geotechnical and environmental engineering firm with hundreds of employees in seven offices.

By the mid-1960s, Chen's firm had been commissioned to investigate and develop geotechnical design criteria for all of Colorado's major college campuses, and even those for several smaller, rural institutions like Northeastern Junior College in Sterling, Colorado. But the company really hit its stride in the 1970s, doing endless soils investigations—several thousand each year—along the Rocky Mountains. Most of the high-rise foundations built in downtown Denver since 1970 have been engineered according to Chen's design parameters.

Fine Arts Building,
Northeastern Junior College,
Sterling, Colorado

One of countless institutional sites for which Fu Hua Chen investigated geotechnical features and developed criteria for subsurface master planning.

Photo credit: Richard Weingardt Consultants, Inc.

Chen himself became active in professional society activities and blossomed as a public speaker and writer. He wrote papers for a wide array of technical journals and general readership publications. His favorite subjects included expansive soils—a common problem soil type in the Denver area—and litigation practices.

Fu Hua was also the author of several books. His most famous, *Foundations and Expansive Soils*, was a fundamental compendium of information on practical design techniques for expansive soils. Translated into four languages and used as a reference and textbook worldwide, many call it the "bible" for professionals who design for expansive soil conditions.

In 1987, he cowrote *Engineering Colorado* (history of consulting engineering in Colorado) with Richard Weingardt. His 1992 book *Between East and West* is a spellbinding account of his career from being chief engineer on the Burma Road to completing noteworthy projects successfully in the United States. Active until the end, Chen had a major manuscript completed and ready for publication when he passed away in 1999 at age 87.

Fu Hua served as a great mentor and teacher, not only to young engineers in his firm, but also to emerging leaders throughout the industry.

He was the only engineer to have been elected president of all four main engineering groups in Colorado: ASCE, National Society of Professional Engineers (NSPE), American Council of Engineering Companies (ACEC), and Colorado Engineers Council (CEC). He also served as senior vice president of ACEC nationally and on influential civic committees appointed by the governor.

In 1980, Chen received an honorary doctorate of science degree from Colorado State University, and, in 1982, he received the prestigious Gold Medal from the Colorado Engineering Council (an association of all of the state's engineering societies), in recognition of his amazing career and wide-reaching impact on the profession.

During the liability nightmare times of the 1980s, Chen was instrumental in getting tort reform legislation enacted in Colorado. Among these reforms were bills curtailing frivolous lawsuits and shortening the length of time engineers could be held liable for past actions.

Chen often told anecdotes about the follies of greed including this one: "How do you catch a monkey? You cut two small, hand-sized holes in a hollow tree and put honey in the tree. When the monkey reaches in to grab the honey, its hand becomes a fist that will not allow passage out of the hole. The monkey either has to drop its prize or hold on and stay handcuffed to the tree."

Having great wealth, affirmed Fu Hua, does *not* translate into being happy. He said, "The secret of happiness is hard work. When you're assigned a job, you do it no matter how difficult. If you're given everything, it makes you weak. If you work for it, it makes you strong" (Dumas 1996).

In the late 1980s, Chen and Associates was sold to a national environmental and geotechnical firm, now part of Maxim Technologies, Inc.

Countless structures all over the Rocky Mountain region and beyond—from single dwellings to skyscrapers, from long-span bridges to dams—rest firmly on foundation systems designed in accordance with requirements set forth by Chen. They stand as a testimony to the wide reach and impact he had on the built environment in the nation's west.

> *"The secret of happiness is hard work. When you're assigned a job, you do it no matter how difficult. If you're given everything, it makes you weak. If you work for it, it makes you strong."*
>
> FU HUA CHEN

CHAPTER THREE

Transportation Trendsetters

*The emergence of effective transportation
during the last half of the 19th century and
the 20th century has been a large factor in
the political, economic, and social revolution
which has produced our dynamic society.*
—Richard Kirby

As with early canal building, the engineering of colonial America's first major road-building projects was greatly influenced by European civil engineering practices, and the British civil engineer having the singly most significant sway on American's early road construction practices was John McAdam (1756–1836).

After spending much of his youth in America, the 28-year-old McAdam returned to England in 1783 at the conclusion of the Revolutionary War and immediately began making his mark in the "Mother Country." His prowess at road building resulted in his appointment, in 1827, as the surveyor general of all the roads in the United Kingdom.

His revolutionary method, known as "macadamizing," produced superior and highly durable roadways. It consisted of a base course of large stones laid on compacted soil footing, a middle course of smaller stones, and then a top course of graded gravel. Gutters along the roadway carried away rainwater and snowmelt. By the middle of the nineteenth century, most main roads in Europe were being constructed using his method, and the United States followed suit (Kirby 1956).

McAdam wrote several widely read textbooks about his engineering methods and became an enthusiastic promoter of road building as an honorable emerging profession. Many bright young Americans heeded his advice, and soon an expanding United States had a solid core of civil engineers highly skilled at producing first-rate highways. Their engineering ability at building highways and railroads, bridges and tunnels, and ports and harbors was attracting serious attention worldwide by the start of the twentieth century.

Equally as impressive was the country's entry into aviation and air travel. After the Wright brothers made their first flight in an airplane in 1903, imaginations ran wild with the possibilities and a budding new transportation field—powered air travel—was born. Civil engineers would play a major role in this field.

The year 1904 saw the emergence of another engineering marvel by U.S. civil engineers—building the massive Panama Canal. When President Teddy Roosevelt had the country take on this seemingly impossible task after the French had failed so miserably at it years earlier, world attention focused on whether the American civil engineering community could meet the challenge. When the United States successfully completed the project in 1914 under the stellar leadership of American civil engineers George Goethals (1858–1928) and John Stevens (1853–1943), world admiration of American engineers and constructors reached an all-time high. American civil engineering had arrived!

After World War II, the skill of America's transportation engineers reached an even higher global standing with the inception of President Dwight Eisenhower's grand plan for a spiderweb of superhighways spanning the country, from coast to coast, and from Mexico to Canada. By the time the U.S. Interstate Highway System was completed in the latter part of the twentieth century, it was the envy of the world—the greatest engineering and construction achievement in history.

Wrote Dan McNichol in his book *The Roads That Built America:*

> We are the only country in the world with such a vast network of superhighways. Not Russia nor China nor all the European countries combined can match it. It is the largest single engineering and construction project on this planet. In scale it is far larger than the Great Pyramids of Egypt, the Great Wall of China, the aqueducts of Rome, and the Suez and Panama canals. But America is often dismissive of her grandest works. When greatness comes often, it is easy to be indifferent about your accomplishments. The Interstate System is a prime example of greatness taken for granted, of overachievement as the norm (McNichol 2004).

The country's skill at engineering and constructing world-class highways, railroads, bridges, tunnels, airports, docks, and harbors—and massive material conveying systems like large pipelines—was extended into space travel in the late 1960s. It was brought about by President John Kennedy's challenge to the nation to put an American on the moon—and beat the Russians by getting there first—before the 1960s ended.

The nation's engineering community quickly and enthusiastically accepted the challenge and rose to the occasion. And it was with a great deal of pride for the profession that, once the engineering feat of the century was finally consummated in July 1969, it was an American engineer—Neil Armstrong—who first put his footprint on the moon's surface. The words he spoke, "That's one small step for man; one giant leap for mankind," are etched in history forever.

An often-overlooked engineering superstar even more key, in many ways, to the success of the moon landing and NASA's entire manned space flight program than Armstrong was civil engineer John Houbolt. Without his solution on how to land on the moon—a major departure from conventional wisdom at the time—President Kennedy's lofty goal would not have been accomplished (Hansen 1995).

In a 2003 article in *The News-Gazette*, Paul Wood reported that NASA had been concentrating on a manned rocket that flew directly to the moon, landed backward (just like they did in the science fiction movies), and took off again. Houbolt said, "It just didn't make sense. For one thing, nobody has ever figured out how to land backward." Wood wrote, "But by orbiting the moon and sending a small landing module to the surface, there was no need for a launching pad, or the enormous amount of fuel to fire a full-sized rocket. It seems obvious now, but Houbolt said people then thought he was crazy" (Wood 2003).

Born in 1919, Houbolt grew up on a farm and attended the University of Illinois at Urbana, earning a bachelor's degree in civil engineering in 1940 and a master's degree two years later. He later earned a doctorate. In addition to contributing immensely to NASA's space program and moon landing, Houbolt came up with solutions for the agency's ever-present problem of aircraft wing flutter on high-speed planes. He also helped design stealth fighters and bombers.

After World War II, he had a hand in bringing German space scientists to the United States, notably the distinguished Wernher von Braun, which bolstered NASA's military-industrial aeronautics and space efforts. Once at the top, von Braun began suggesting that the United States send up satellites. Houbolt, who eventually became good friends with von Braun, said, "Wernher's idea was rejected by politicians. When the Soviet Union launched *Sputnik* into orbit in October 1957, that changed immediately, and we hurriedly geared up for satellites."

According to Houbolt, "At NASA, it was von Braun's word that mattered to me." The two were together in the NASA Mission Control Center in Houston watching the moon landing, and as Armstrong descended the ladder, von Braun turned to Houbolt and said, "Thank you, John." Von Braun credited Houbolt with "saving the space program tens of billions of dollars" (Wood 2003).

Countless other civil engineers who were transportation trendsetters during these times are, like Houbolt, often overlooked, even though their work has been crucial to the world power status of the United States. The four legends featured in this chapter are illustrative of these accomplishments. Three are from America's formative years—Octave Chanute, John Greiner, and Clifford Holland—and one is from the years shortly after the admission of the nation's forty-ninth state—Roy Peratrovich, Jr., a Native American from Alaska.

Octave Alexandre Chanute

Known worldwide as one of the fathers of aviation, Chanute was already a celebrated civil engineer by the time his interests turned to aeronautics. Wilbur Wright said of his fellow aviation pioneer and mentor, "If he [Chanute] had not lived, the entire history of progress in flying would have been other than it has been."

Octave Chanute

Photo credit: Western Society of Engineers

The son of Joseph and Eliza (De Bonnaire) Chanute, Octave was born in Paris in 1832. When he was six, his family immigrated to the United States after his father accepted a position as vice president and history professor at Jefferson College (near New Orleans, Louisiana).

The Chanutes moved to New York in 1846 and, three years later, a 17-year-old Octave was found working at the Hudson River Railroad, where he learned the intricacies of civil engineering as they apply to railroads. This self-taught engineer often said his Hudson River job was the only position he actually applied for in his entire life—he was continuously employed from that moment on without ever again having to seek work.

In 1857, at age 25, Octave married Annie Riddle James. They immediately set their sights westward and within a few years they left New York for Chicago.

By 1863, Chanute had the lofty position of chief engineer for the Chicago and Alton Railroad, involved in engineering and building railroads in Illinois and adjacent states. These assignments allowed Chanute to meet prominent and influential businessmen who commissioned him to design the first railroad bridge to span the mighty Missouri River near Kansas City. During this time, the Union Stock Yards of Chicago selected his design proposal for their new facilities and he became the chief engineer of the yards. He meshed the supervision of its construction with his railroad duties.

Later, returning to the East Coast, Chanute engineered a system to help resolve New York City's rapid transit problems of the day. His system replaced horse-driven cars with elevated railroads powered by steam locomotives.

His novel designs and experience building complex bridges and railroad systems—plus his knowledge of materials—made him a giant in the industry by the late 1870s. His work on preservation of materials led him to developing procedures for pressure-treating rail ties and telephone poles with creosote and other preservatives. Several of his pressure treatment techniques continue in use throughout the world today.

Having attained an outstanding reputation and financial success as a civil engineer, Chanute began to make plans to retire. In 1875, he visited Europe and learned of extensive efforts underway there—especially by F. H. Wenham in England—to develop manned flight.

In 1890, at age 58, he returned to Chicago and retired from his first career. Then he began intensely pursuing his interest in aviation, corresponding with engineers and aviation pioneers internationally and gathering all the leading-edge information known at the time. This culminated in a series of papers titled "Progress in Flying Machines," first published in trade journals and later in a highly respected book with the same title.

Considered the classic book on early aviation, the 1894 *Progress in Flying Machines* gave the world its first compendium of flight. This earned Chanute the title of the first "aerohistorian."

> *"If Octave Chanute had not lived, the entire history of progress in flying would have been other than it has been."*
>
> WILBUR WRIGHT

Chanute did not confine his research to reading and writing about prevailing theories and doing laboratory experiments. He actually built and flew full-scale gliders, making hundreds of flights at the sand dunes near present-day Gary, Indiana. These efforts allowed him to perfect his bi-plane glider design to the point of recommending its use to develop a powered flying machine. Chanute's bi-wing glider became the structural model for the Wright brothers' first successful airplane, the "Flyer."

During this time, Chanute also organized several conferences to bring together world-class experts to discuss the latest findings and advances in aviation, including a highly successful International Conference on Aerial Navigation at the World Columbian Exposition in Chicago in 1893—the same world's fair that had George Ferris's fantastic wheel as its main attraction.

After learning of Chanute's forward-looking activities and successful glider flights and reading his papers on flying machines, the Wrights—Wilbur (1867–1912) and Orville (1871–1948), both greatly impressed with Chanute's vast knowledge of aeronautics—began a lifelong friendship with him.

In 1901, Chanute visited the Wright brothers and encouraged them in their gliding experiments. He was profoundly stirred by their ambition to build an airplane, and he freely shared all the knowledge he had accumulated on aviation and aeronautics over the years.

Even before their first face-to-face meeting, the two brothers corresponded with Chanute on a routine basis, spelling out their thoughts on a powered airplane in careful detail. Although he witnessed many of their progressively aggressive glider flights, he was not present at Kitty Hawk, North Carolina, on December 17, 1903, when the Wright's "Flyer" lifted off and flew, to the total amazement of hard-core skeptics of airplanes. More than anyone alive, Chanute knew what this accomplishment would mean and how travel and the future would be impacted.

Nearly 10 years earlier, in his classic 1894 book on flying, Octave had written, "Let us hope that the advent of a successful flying machine, now only dimly foreseen and nevertheless though to be possible, will bring nothing but good into the world; that it shall abridge distance, make parts of

the globe accessible, bring men into closer relation with each other, advance civilization, and hasten the promised era in which there shall be nothing but peace and goodwill among all men" (Chanute 1976).

The 1903 Wright Flyer, Orville and Wilber's first flight at Kitty Hawk

Octave Chanute's perfected bi-wing glider design was the model for the Wright airplane.

Photo credit: National Air and Space Museum, Smithsonian Institution

Colonel Warren Roberts, one of Chanute's protégés, wrote, "At the time I made his acquaintance in 1892, he was very enthusiastic about the possibility of man flying. He endeavored to interest young engineers in the subject" (WSE 1992).

Commenting on Chanute's mentoring style, Roberts reported, "He never seemed to be in hurry and was so approachable that a young engineer readily made his acquaintance. He always appeared to be happy to take the time to explain clearly any problem we presented to him."

About his overall knowledge, Roberts said, "He impressed me as having read widely and wisely. He seemed like a walking library to a young engineer—an engineering authority. Octave was surely the most versatile engineer I have ever met. He had the most multitalented mind of any American since Thomas Jefferson."

Although the recipient of many honors and accolades—the town of Chanute, Kansas, for instance, was named after Octave, as was Chanute Air Force Base at Rantoul, Illinois—Chanute stood out not for what he received but for what he gave. He generously gave his time and support of others, and the engineering profession as a whole, including being the 1891 president of ASCE.

Shortly after he had retired as the president of the Western Society of Engineers (WSE) in 1901, he endowed a WSE fund—the Octave Chanute Prize—to be given annually to any WSE member whose engineering paper is judged outstanding and/or forward-thinking.

Chanute always believed in sharing knowledge. For him, technical information was a public commodity, never to be hidden in secrecy. The same analytical persistence that made him a successful civil engineer made him an outstanding aeronautical authority.

Octave passed away on November 23, 1910, at age 78. At his funeral, Wilbur, the older and more outgoing of the two famous Wrights, gave the eulogy and explicitly acknowledged that the great civil engineer's labors "had vast influence in bringing about the era of human flight."

John Edwin Greiner

On May 1, 1908, after years of distinguished service with the Baltimore and Ohio (B&O) Railroad, 49-year-old John Greiner resigned his position as assistant chief engineer and established a private consulting engineering practice—John E. Greiner Consulting Engineers. Thus began a legacy of a one-man firm that grew into and became part of one of the world's largest engineering companies—URS Corporation, which in 2000, employed more than 15,000 people in 28 countries worldwide.

Born in Wilmington, Delaware, on February 24, 1859, to Annie Steck and John Greiner, the sexton of Zion Lutheran Church in Wilmington, young John grew up on his family's farm. There, he and his brother experienced a Tom Sawyer–type boyhood. They roamed the countryside, hunted and fished, and milked cows and helped with farm chores, like many other American-born youngsters in rural America of the late 1800s (Tiedeman 2001).

Their mother, an educated and musically accomplished woman, introduced her sons to the finer points of music early on. John studied the violin and performed as a member of the local Beethoven Musical Club while still in secondary school. His zeal for music—and enjoyment of outdoor sports—stayed with him all his life.

After graduating with a civil engineering degree from Delaware College, which he attended on a legislative scholarship, young Greiner worked for the Edgewater Bridge Works as a draftsman. In 1884, he was employed by the Keystone Bridge Works in Pittsburgh to make engineering calculations, prepare detailed shop drawings, and inspect rolled iron and steel members at various company mills. He was then hired by the Austrian-born bridge builder Gustav Lindenthal (1850–1935) as his resident engineer during the erection of the Seventh Street Bridge over the Allegheny River.

Educated at the Polytechnical Institute in Dresden, Germany, Lindenthal quickly established himself as one of America's greatest bridge builders after arriving in 1874. The Seventh Street Bridge was one of his first two major bridges, the other being the 1883 replacement of John Roebling's 1845 suspension bridge over the Monongahela River (Billington 1983).

As the designer of many of America's greatest bridges in the late 1880s and early 1900s, Lindenthal is often compared to the great Roebling. In addition to Greiner, Lindenthal became a mentor to many of America's bridge-building icons of the early to mid-1900s. On his Hell Gate Bridge project (completed during World War I) alone, David Steinman and Othmar Ammann cut their big-bridge-designing teeth while in his employment.

In 1885, Greiner began his 21-year career with the B&O Railroad Company, where he progressed from draftsman to chief draftsman and bridge inspector, then to engineer of bridges and buildings, and, finally to assistant chief engineer.

One year after joining B&O, in December 1886, he married Lily Flower Burchell and the union produced two daughters, Lillian and Gladys. Not long after his wedding, John became quite active in his community. Appointed by Baltimore Mayor Robert McClean, he served on the city's commission to deal with the devastation of the 1904 fire that gutted downtown Baltimore. Shortly after, McClean named Greiner to a commission for revising the city's building code (Tiedeman 2001).

John Greiner

Photo credit: University of Delaware Archives

While with B&O, Greiner patented a type of bridge used to replace a large number of deteriorated wooden highway bridges crossing B&O tracks. Realizing that the railroad yards were filled with old rails that had no market value, he used them to rebuild the worn-out wooden bridges. Since the rails were basically all the same size, his patented bridge system was designed with member stresses as uniform as possible, resulting in structures that could be built at much lower prices than timber replacements would cost.

By 1901, Greiner had a national reputation for developing railroad bridge specifications. His standards for rating the load-carrying capacities of bridges were adopted by the American Railway Engineering Association in 1908. Because of his expertise, he was often asked to examine and rate bridges nationwide, including the first major steel bridge in America—St. Louis Eads Bridge.

Once he established his private consulting business in 1908, success came rapidly to Greiner. He designed numerous bridges for the Richmond, Fredericksburg, and Potomac Railroad Company, including the million-dollar James River Bridge in 1916 and the Rappahannock River Bridge in 1924. During this time, he received another degree from the University of Delaware—an Sc.D. in 1917.

Right after World War I, the United States earnestly began expanding its national transportation system. To overcome deficiencies in public funds to finance these projects, substantial private funds were raised by local companies. These companies were formed specifically to finance bridge construction and then operate them as privately owned toll facilities. Greiner's company played a prominent role in engineering many of these crossings over major rivers, including the Ohio, James, Potomac, Delaware, and Susquehanna.

In addition to his highway and bridge activities, Greiner was chairman of the Port Development Commission of Baltimore from 1920 to 1927, supervising a $50-million improvement to the city's port. In 1928, he received a gubernatorial appointment to the powerful Maryland State Waterfront Commission.

The beginning of the 1930s saw a reduction in Greiner's corporate staff due to the Great Depression. The company's strength and resiliency, how-

Denver International Airport

The firm John Greiner founded was the City of Denver's construction manager for the building of its $5 billion, 53-square mile facility. When completed in 1995, DIA was the country's largest public-works project up to that date. It's the only airport in the world capable of landing three parallel aircraft simultaneously during severe weather conditions.

Photo credit: Richard Weingardt Consultants, Inc.

ever, showed itself in the late 1930s when the country began its economic recovery. The Greiner Company's growth and prosperity during this time—and up until (and even through) World War II—matched the best in America, in large part due to its founder's well-established reputation within bridge design circles.

Many of the firm's noteworthy projects during this time and into the early 1940s were major bridges and highways leading to and from them. One of the more prominent of these projects was the 1940 Thomas J. Hatem Memorial Bridge along Maryland State Highway 40 at Havre de Grace in Hartford County. This spectacular multispan, arched steel truss bridge, which extends for 7,618 feet over the Susquehanna River where it meets the Chesapeake Bay, rides high above the water—177 feet above at its highest point.

Even though Greiner passed away shortly after the Japanese attacked Pearl Harbor and the United States entered World War II, the firm continued on, expanding what he had started and capitalizing on his long list of financially successful projects, especially in transportation. In the country's postwar growth and the surface transportation boom years of the mid-to-late 1900s, Greiner's firm rapidly expanded its highway design expertise. Because of this decision, the company's growth closely paralleled the expansion of America's surface transportation network, in particular the interstate highway system.

After American Airlines instituted same-day passenger air service from New York to Los Angeles on January 25, 1959, the budding commercial airline industry in the United States began demanding better and safer airports and the firm Greiner founded—following his core philosophy to always be at the forefront of technological advancement—immediately expanded its cutting-edge transportation know-how into that field.

By the early 1990s, the company's world-class reputation in airport design was such that Greiner was selected, after an extensive national search, to be the city of Denver's overall construction manager for the design and construction of its massive, totally new, and innovative airport—Denver International Airport—which opened in 1995 at a cost of $5 billion.

Although John Greiner stayed active in his firm well into his late seventies, he stepped down from his day-to-day duties as chief executive officer when he hit 80. He died suddenly in 1942 after a brief illness. He was 83 years old.

The overwhelming international response to his death reflected the widespread appreciation and recognition many had for the energy and dedication he put into making both his profession and his community better. His leadership had influenced so many people that, at his funeral, several hundreds of telegrams and letters flooded in from all over the world—a testimony to the impact of Greiner's remarkable career.

Throughout his professional life, Greiner received many accolades, including the ASCE Norman Medal in 1896 and the James Laurie Prize in 1915. His paper "What Is the Life of an Iron Railroad Bridge?" was the basis for his winning the Norman award. He was elevated to Honorary Member of ASCE in 1932 and, in 1984, his name was added to the University of Delaware's exalted Alumni Wall of Fame.

In the mid-1990s, the company joined with Woodward-Clyde and, shortly after, merged with URS. The combined firms are now known as URS Corporation, one of the largest engineering companies in the world.

> "*Our railroad yards are filled with discarded old rails (all basically the same size), let's engineer some structural systems—with uniform member stresses—that allow the reuse of these rails as structural elements to produce low cost bridges.*"
>
> JOHN GREINER

Clifford Milburn Holland

When 41-year-old Clifford Holland died of exhaustion and heart failure on October 27, 1924—before his great tunnel under the Hudson River was finished—he was eulogized in the press as the "martyr engineer" for his heroics and dedication. He was also acclaimed "the most noted tunnel builder in the world" and the engineer of the "eighth wonder of the world"—the Holland Tunnel, linking downtown Manhattan with New Jersey.

The project was named in his honor within two weeks of his passing. In 1999, *Engineering News-Record* (ENR) honored him as one of the 10 most outstanding "Landmark Project Engineers" of the last 125 years.

In giving Holland's own perspective about his life, Stuart McCarrell, author of the New York City drama *Voices, Insistent Voices,* had Holland enthusing, "Things broke well for me: the choice of engineering, the good school, the right years—when work was booming, the perfect place——New York, the lucky first job—subway shafts to Brooklyn, then my life's pinnacle and purpose—the great tunnel to New Jersey, and my wife, noble, strong. She knew what the project meant to me, and the city" (McCarrell 1996).

The only son of Edward John and Lydia Francis (Hood) Holland, Clifford was born in Somerset, Massachusetts, on March 13, 1883. He was the great grandson of Somerset's renowned Deacon Nathan Davis, a descendant of Francis Le Baron of Plymouth and Roger Williams of Providence. He attended local Somerset public schools, graduating from Fall River High School. The engineering bug bit him early; he told schoolmates he was going to be a "tunnel man" while still in his preteens (Somerset 1974).

Clifford Holland

Photo credit: Vincent Tocco, Jr.

He entered Harvard University's engineering program in 1902 after graduating from Cambridge Latin School. To help pay college expenses, he tutored classes in evenings, waited on tables at the college dining hall, and worked as a gas meter reader for the city in the summer months. From Harvard, he earned his B.A. in 1905 and B.S. in civil engineering in 1906.

During his last year at Harvard, he passed the New York State civil service examination and, upon graduation, took a position as assistant engineer with the Rapid Transit Commission of New York, designing and building subways and tunnels. His most significant assignment early on included the construction of old Battery Tunnel, where his indoctrination into the unique intricacies of urban tunnel construction took shape.

When he married Anna Coolidge Davenport of Watertown, Massachusetts, on November 5, 1908, the bespectacled and scholarly looking Clifford was already a notably serious engineer and unquestionably a "young man on the rise." The couple would have four daughters.

In 1914, the 31-year-old Holland was promoted to the position of tunnel engineer for the commission, in full charge of design and construction of four double subway tunnels under the East River. (Contract value of the work was $26 million at the time.) He was elevated to division engineer in 1916, a title he held until 1919, when he was hired as chief engineer to build the great tunnel for the New York State and New Jersey Interstate Bridge and Tunnel agencies. By then, Holland had an international reputation as one of most outstanding experts and leaders in the field of subaqueous construction.

The Holland Tunnel was commissioned out of desperation. By the early 1900s, the ferries of the Hudson River—the only mode of travel between New York City and New Jersey—were tremendously overloaded, especially after the proliferation of Ford's hot-selling 1908 Model T. Thousands of cars and trucks and legions of horse-driven buggies and wagons crowded the riverbanks, causing mind-boggling congestion and traffic jams every day.

After years of study, engineers and the powers that be determined that a major tunnel rather than a bridge was the answer. It would double the traffic capacity across the river and cut the time ferries took by more than half. But a modern vehicular tunnel of the proportions and complexity required had never been attempted. Was it even possible? How could it be done without asphyxiating its users? Without some way of eliminating all the poisonous carbon monoxide from the automobiles in the tunnel, drivers could pass out before reaching the other side!

A number of prominent engineers, including George Goethals of Panama Canal fame, submitted tunnel design proposals to accomplish the feat in 1919. The design selected came from one of the "Big Apple's" own, the confident and unstoppable 36-year-old Clifford Holland. But his design would require creating the first bona fide mechanical ventilation system ever for a tunnel. And he had his critics.

Because a vehicular tunnel of this type had never before been completed, many of the engineering problems involved were unprecedented. Holland's tunnel scheme—actually two separate 30-foot diameter tunnels (tubes) running about 50 feet apart through the silty riverbed, 94 feet below the Hudson's surface—stretched over a mile and a half. The two giant tubes were divided into three horizontal layers—the middle one to handle traffic and the top and bottom ones to move air. To accommodate two lanes of one-way traffic, the roadway width in each tube was 20 feet.

> *"Things broke well for me: the choice of engineering, the good school, the perfect place—New York, then my life's pinnacle and purpose— the great tunnel to New Jersey."*
>
> CLIFFORD HOLLAND

Four large ventilation towers, two on each side of the river, housed 84 fans, each eight feet in diameter. Half of them blew clean air into the lowest space in each tunnel, forcing it into the roadway through narrow slots along the curb. The other 42 extracted dirty air from the roadway through roof vents into the upper space and eventually out to the towers, where it was discharged. With this design, the air in each tunnel changed every 90 seconds at a rate of 3.6 million cubic feet per minute (Gray 1927).

Work began on Holland's grand passageway in 1920, with crews working east from Jersey City and west from lower Manhattan. On October 29, 1924, two days after Holland died, the two teams met each other, not inches off line. The accuracy of their "holing through" set the standard for future underwater tunnels, including the $18-billion Chunnel under the English Channel completed seven decades later, using much more sophisticated guidance control systems.

When the $48-million Holland Tunnel opened to traffic in 1927, two years before the great stock market collapse of 1929, the toll cost 50 cents and the trip took eight minutes. President Calvin Coolidge officially opened the tunnel by signaling from the presidential yacht, *Mayflower*. Then on November 12, 20,000-plus people walked down its white tiled corridor and 51,694 vehicles drove through, with a Bloomingdale's department store truck being the first.

Holland Tunnel, entrance on the New York side, 2004

On opening day in 1927, thousands New Yorkers crowded into—and/or drove through—the more than 8,500 foot long tunnel, catching a quick glimpse of "the eighth wonder in the world," an underground project unprecedented in size and sophistication, equipped with the world's first-ever transverse-flow ventilating air system. The Holland became the model for all subsequent underwater tunnels.

Photo credit: Richard Weingardt

Nearly two billion vehicles have passed through it since opening; today, more than 100,000 vehicles use the passageway daily.

The energetic Holland was active in many engineering societies, in particular, the ASCE (where he served on its Board of Direction), the American Association of Engineers, and the Harvard Engineering Society (where he was president). In his honor, the engineering scholarship of the Harvard Society was renamed the Clifford M. Holland Memorial Aid in Engineering Scholarship.

Holland's legendary devotion to excellence and duty, along with his 'round-the-clock efforts during construction, made possible the successful and timely completion of his engineering masterpiece—and, unfortunately, hastened his own physical deterioration.

The Holland Tunnel, in its time, set a precedent for the construction of tunnels. It remains the model for major underwater passages even today.

Roy Peratrovich, Jr.

In 1962, three years after the state of Alaska was admitted to the United States, Roy Peratrovich—a member of the Tlingit Indian Tribe, Raven Clan and Raven Family (double Raven)—became the first Alaska native to receive a license as a registered professional civil engineer in the state.

Roy was born in May, 1934, in Klawock, a small southeast Alaska native fishing village. He was the eldest of three children of Roy and Elizabeth Peratrovich, both civil rights leaders credited with the passage of the first antidiscrimination law in the United States in 1945. Roy Senior's father, John, an immigrant from Yugoslavia, established the first Alaskan salmon cannery at Klawock. Elizabeth's father, Andrew Wanamaker, was a Presbyterian missionary.

Young Roy's education began in a one-room schoolhouse in Klawock before his family moved, in 1941, to the territorial capital of Juneau. There, he was among the first Alaskan natives to attend public school. In 1952, the Peratrovich family moved to Denver, Colorado, where Roy lettered in football and graduated from South High School in 1953. He earned his bachelor's degree in civil engineering at the University of Washington in 1957 and became a proud member of Tau Kappa Epsilon fraternity. After graduation, he married and immediately went to work as assistant bridge design engineer for the city of Seattle, designing bridges, overpasses, and interchanges.

In 1961, Peratrovich returned with his young family to Juneau to take an assignment as one of three squad leaders in the Alaska Department of Highways Bridge Design Section. "Our Bridge Design office was located in an old abandoned building that had been part of the Alaska-Juneau Gold Mine," recalled Roy in a 2002 interview. "The state had just inherited all the bridges and highways (such as they were) from the federal government. We didn't know where they were located, how many there were, or what condition they were in, so we created a filing system filled with all available information so it could be used in the future. We developed the first detailed system for doing annual bridge inspections, as well as the evaluation process and condition rating of existing bridges statewide. And Alaska is a big state!"

Roy Peratrovich, Jr. with one of his famous sea otter sculptures

Photo credit: Roy Peratrovich, Jr.

Much of the bridge design Peratrovich (and his colleagues) did was revolutionary. For example, he designed the first modern all-steel, all-welded bridge constructed in Alaska—the Cordova ferry transfer bridge. Its many innovative design concepts became state standards.

Another leading-edge design, considering the state's erratic ice conditions and seismic forces, was his 1,000-foot bridge over the ice-ridden Susitna River. It used continuous steel plate girders with clear spans of 250 feet designed to rest on heavy-duty, pile-supported concrete piers.

One of his last assignments before leaving the bridge section was the design of the first cable-stayed bridge in the nation at Sitka, Alaska. A controversial concept in those days, this cable-stayed design was not accepted at first. It was later completed by his future partner, Dennis Nottingham.

In 1969, Peratrovich was promoted to section head for the highway department. At his new position, Roy was in charge of the studies for Alaska's first two major metropolitan transportation systems still in operation in Anchorage and Fairbanks.

Peratrovich left the government in 1972 and joined a private engineering firm in Anchorage to work on the Trans-Alaska Pipeline. He was involved in the design of two highline structures—spanning 1,000 feet and 1,500 feet respectively—to carry the four-foot-diameter oil pipeline on a slope too steep to stand on. Roy also designed the 400-foot tied arch pipeline bridge that spans the Gulkana River, a designated wilderness river underlain in permafrost.

In building this bridge, Peratrovich and his fellow engineers introduced the use of driven steel piles in permafrost, which became a preferred construction method in the Arctic. This foundation system became the prototype for all structures along the entire length of the massive $8-billion pipeline project.

Said Peratrovich, "Anyone working on the Trans-Alaska Pipeline quickly learned what true engineering is all about, and then some. Major advances in engineering were made almost on a daily basis." The ice force criteria and design concepts his team worked out were implemented on almost every bridge along the Pipeline.

> "*Anyone working on the Trans-Alaska Pipeline project quickly learned what true engineering is all about, and then some. Major advances in engineering were made almost on a daily basis.*"
>
> ROY PERATROVICH, JR.

The history-making, one-of-its-kind undertaking took three years of round-the-clock activity to complete. Engineering design work and final construction had to function in and withstand Arctic conditions, frozen soils, and temperatures ranging from 100 degrees Fahrenheit to minus 60 degrees as well as 100-plus mile-per-hour winds, major earthquakes, and the force of ice floes as thick as seven feet.

After the successful completion of the Pipeline in 1977, Peratrovich accepted a position in Juneau with the newly formed Alaska Department of Transportation and Public Facilities. He headed a division created to develop uniform procurement policies and procedures for designing and constructing all public facilities in Alaska.

Two years later, in 1979, he founded his own consulting engineering firm, Peratrovich Consultants in Seattle, to work on the design of the massive West Seattle Bridge. His firm coordinated and scheduled 20 other consultants on the project.

Shortly after, Peratrovich and his long-time colleague Dennis Nottingham established the firm of Peratrovich and Nottingham in Anchorage. Now named Peratrovich, Nottingham and Drage (PN&D), it has offices in three states, with engineering projects across the United States and overseas.

Through the years, PN&D earned numerous engineering excellence design awards, including a record 40-plus James F. Lincoln Arc Welding Awards. Two of the many award-winning innovations created by PN&D during Peratrovich's tenure were:

- The Spin Fin Pile: Heavy-walled steel pipe piles with steel fins welded at a slight angle to the pile tip. As the pile is driven, it spins into the ground. The anti-unscrewing capacity developed creates a tremendous uplift and bearing strength, double that of a smooth pile.

- The Gunderboom: An environmental curtain for use in oceans and rivers to contain oil spills and offshore marine construction debris, control siltation, and protect swimmers from contamination. Its unique geotextile net is held in position with marine anchors. The system was successfully used to contain damage after the *Exxon Valdez* oil spill tragedy in 1989.

The Gulkana River Bridge on the Trans-Alaska Pipeline

The Gulkana's foundation system became the prototype for all structural foundations on the $8-billion Pipeline, which took three years to build.

Photo credit: Peratrovich, Nottingham and Drage, Inc.

Active in the business and engineering communities throughout his career, Peratrovich helped found the Architects and Engineers Insurance Company (AEIC), a professional liability insurance company for architects and engineers, in the troubled 1980s when it was impossible for many engineering firms to obtain errors and omissions insurance at any price. Beginning in 1987, he served on AEIC's Board of Directors for 12 years.

Peratrovich has held leadership roles on numerous public boards and commissions, including the Alaska Board of Architects, Engineers and Surveyors, Alaska Visual Arts Board, Alaska State Board of Welding Examiners, City and Borough of Juneau Planning and Zoning Commission, Juneau Recreation and Planning Committee, and Juneau Road Standard Committee.

In 1991, he left his home state for good, transferring to PN&D's Seattle office. Eight years later, the 65-year-old Native American retired from the firm he helped establish, taking up a lifelong dream to create sculptures. He founded Ravenworks Art Studio, and today many of his bronze sculptures grace corporate offices and private homes around the country. His bronze bust of his famous mother, Elizabeth, resides in the rotunda of the capital building in Juneau.

This bold and creative civil engineer-turned-artist currently lives on Bainbridge Island near Seattle with his wife, Toby, and a menagerie of pets. He maintains close ties to Alaska through his siblings, three grown children, and one grandson, all of whom live in Anchorage.

CHAPTER FOUR

Builders of Bridges

> *There can be little doubt that in many ways the story of bridge building is the story of civilization. By it we can readily measure an important part of a people's progress.*
> —Franklin D. Roosevelt

Bridges are national treasures and among the most wondrous things humankind has ever engineered. Because they are highly visible, nothing better symbolizes the beauty of civil engineering and the infrastructure of the built-environment than bridges. Each bridge is unique, reflecting use of topographical constraints as well as the talents of its design engineer and builder.

Designed to carry people, vehicles, and materials over an infinite variety of artificial and natural barriers, their singleness of purpose distinguishes them from other forms of structure. The design of buildings and "people" structures, for example, is concerned with people requirements: heating, ventilation, lighting, space requirements, and so on. Most of the time, the structure itself is not the dominant factor. But in a bridge, the structure dominates—exposed for all to see.

The public is fascinated by and has a universal affection for bridges. In America, nearly 600,000 bridges span gorges, rivers, and other obstacles across the country, ranging from small timber structures in rural areas to majestic, towering soaring landmarks such as the Golden Gate Bridge (1937) and the Brooklyn Bridge (1883), both of which were world record-holder suspension bridges when built.

Bridges are perhaps the most critical components of the surface transportation network because they ensure the continuity of mobility. They are a major public works investment and a critical link in the transportation network. The chain is only as strong as its weakest link; thus, roads and highways are only as sound as the bridges that connect them.

Bridges are also important because of their usefulness. To be sure, as a vital aspect of their design, their visible appearance should not be discounted. But, at the root, bridges serve practical, utilitarian functions, which is the essence of their existence. Bridges become part of our landscape, both in beauty and in function. They have become points of national pride for every country around the world.

Building bridges represents the "can-do" spirit of America. And the four legendary builders of bridges spotlighted in this chapter are indeed vibrant examples of that spirit. Although they were all born in the middle half of the nineteenth century, an earlier generation of "can-do" bridge builders populated the country well before then. And no better understanding of that generation's daring spirit can be grasped than by briefly looking at those responsible for early America's romantic and picturesque covered bridges of the late 1700s and early 1800s, a colorful chapter in pioneer engineering and construction.

Although covered wooden truss bridges are often seen as being uniquely American, covered bridges per se were not original in America. The first recorded covered bridge spanned the Euphrates River in Babylon in 785 B.C., and Marco Polo's journals mention Chinese bridges having "handsome roofs." Through the ages, Europe had its share of covered bridges. However, most were of the timber trestle type, short spans bearing on stone piers.

The truss structure—a characteristic of U.S. covered bridges—was rarely used in Europe until after it became common in American bridges, even though the truss was the invention of sixteenth century Italian engineer-architect Andrea Palladio. While he actually built a few small bridges incorporating truss designs in the 1570s, Palladio was far ahead of his time, and Europeans who favored the stone arch subsequently neglected his work.

America's first noteworthy covered bridge, constructed in 1805, was a three-span masterpiece of carpentry extending 550 feet over Pennsylvania's Schuylkill River. It was designed and built by Timothy Palmer, a 52-year-old, raw-boned "New England Yankee" master carpenter from Newburyport, Massachusetts. Its heavy timber and arch-trusses were covered with a roof and siding as protection against the weather at the insistence of stubborn Judge Richard Peters, head of the building committee (Griggs 2004).

After Palmer's great bridge was finished, the challenge for erecting bigger and longer bridges quickly attracted a covey of "self-taught" bridge "engineers." Outstanding among them were New Englanders Jonathan Walcott and Theodore Burr, and Lewis Wernwag, a young immigrant "engineer" from Germany (Nautilus 1988).

In 1812, Walcott, who was 32 years old at the time, won a nationwide competition to build the longest covered wooden bridge ever built anywhere in the world. The 30-foot-wide, multispan Kingpost truss structure, spanning Pennsylvania's mighty Susquehanna River, cost $150,000. Unfortunately, it was washed away in a flood less than 20 years after it was built.

Although Burr lost the distinction of engineering the longest multi-span wooden bridge in the world to Walcott, the congenial genius from Connecticut accomplished a far finer achievement. He designed and built the McCall's

Ferry Bridge over the Susquehanna, a covered wooden bridge having the world's longest single span—360 feet long. Plus, he invented the "Burr arch-truss," which became the model for hundreds of early U.S. bridges.

Wernwag was responsible for building the second longest single-span wooden bridge ever built to that point, the spectacular "Colossus" over the Schuylkill. It became the most celebrated of American covered bridges until it burned down in 1838. Its flared kingposts bracing a double arch were the hallmark of Wernwag's many bridges.

Over 10,000 covered wooden bridges were built in the United States between 1805 and 1885, but fewer than 1,000 remain. Although 30-some states still have covered bridges, they are disappearing at an alarming rate, and along with them goes part of early Americana—the mystique of its covered wooden bridges and the memory of the engineers who created them. The four states that have the most covered bridges today are, in order, Pennsylvania, Ohio, Indiana, and Vermont. The oldest existing covered bridge is the Hays Bridge in Pennsylvania, a 110-foot long conventional Burr arch-truss, built in 1825 at a cost of $1,500.

Possibly the most daring "can-do" U.S. bridge builder in the 1870s was Indiana-born James B. Eads (1820–1887), designer of the Eads Bridge over the Mississippi River at St. Louis, Missouri. This 52-year-old self-taught engineer had little if any bridge construction experience at the time. But that did not stop him from designing what turned out to be not only the world's first large structural steel bridge but also the biggest bridge yet built. It was completed in 1874, nine years before the Brooklyn Bridge.

Eads's revolutionary masterpiece, which was fashioned after a railroad bridge in Koblenz, Germany, contained a center arch spanning 520 feet and two side arches of 502 feet. Its tubular steel arches were built using the cantilever method, a first for large-scale bridge construction. It also featured the first U.S. application of deep compressed air caissons in the building of underwater piers. Eads never built another bridge. This bridge, however, has been designated a National Historic Civil Engineering Landmark.

Although Eads's bridge and the structures of the master bridge builders profiled here—Roebling, Modjeski, Waddell, and Steinman—are of steel, iron, or concrete, and considerably less apt to waste away than America's wooden bridges were, they are not indestructible. As national treasures, it behooves the engineering community to make sure they are properly maintained for future generations. Who better to be the stewards of that charge than civil engineering leaders and civil engineering groups?

Emily Warren Roebling

Emily Roebling, the first woman ever to address the American Society of Civil Engineers, was a builder, field engineer, financier, lawyer, and community leader. A heroic nineteenth-century figure, Emily became a role model for women of all ages in all fields.

She was the single most important person responsible for the timely completion of the Brooklyn Bridge, the longest suspension bridge—and mightiest engineering achievement—of its era. Because of her enormous contributions to its success, her name is prominently inscribed (along with those of her husband and father-in-law, Washington and John Roebling) on the Brooklyn's dedication plaques located on each of the bridge's two massive end towers.

Emily Roebling

Photo credit: Rutgers University Libraries, the Roebling Family Papers Collection

Born on September 23, 1843, in the village of Cold Spring in Putnam County, upstate New York, Emily was the eleventh of twelve children, only six of whom survived to adulthood. Her parents were Sylvanus and Phebe (Lickley) Warren. Her father was a New York State assemblyman and supervisor of the town of Philipstown (which included Cold Spring).

It was there, along the Hudson River, that this youngster developed into an enthusiastic and expert horsewoman. It would be a pastime that, along with driving horse-drawn carriages, would endure well into her later years. Eventually, she would also become expert at riding bikes, a pursuit not especially admired by "proper" Victorian women of her day.

Although socially prominent, the Warren family was not especially wealthy. Emily, however, was quite proud of her family history. Later in life, she became an active member of the Daughters of the American Revolution, the Society of Colonial Dames, the Colonial Daughters of the Seventeenth Century, the Holland Dames of America, and the Huguenot Society. Her interest in these hereditary organizations, however, was not grounded in snobbery or basking in the glory of ancestors. Emily (Warren) Roebling was never unwilling to "get her hands dirty" with real work. Rather, her interest was in preserving America's stirring history as an inspiration for future generations (Weigold 1984).

Emily had an extremely close relationship with her "big" brother Gouverneur K. Warren, even though he was 13 years her elder. She adored him for a number of reasons, but especially because he was a Civil War hero, witty, and socially adept. He was both a mentor and a role model for her. A college-educated civil engineer, Gouverneur rose to the level of Major General and Commander of the Fifth Army Corps during the war and was honored several times for his exploits on the field and victories in battle.

It was while visiting her brother during a military dance at his post that Emily met her future husband, Washington Augustus Roebling, the eldest son of America's foremost bridge builder, John Roebling. Washington was a highly skilled and respected, up-and-coming civil engineer on General Warren's staff. And even though he was six years older than Emily, their strong mutual attraction was clearly obvious. The reserved army colonel, however, seemed the more smitten of the two. He found his Major General's young sister both good-looking and "exceptionally delightful."

In a letter to his sister hinting that something serious might be in the works, Washington conveyed this description of Emily: "She is attractive, dark-brown eyed, slightly pug-nosed, lovely mouth and teeth, no dimples in her cheeks, the corners of the mouth supply that. She is a little above medium size and has a most lovely complexion . . . and is a most entertaining talker." Talking was something he thought he was not especially good at, or—along with socializing—comfortable doing.

After a steady courtship that included a constant barrage of intimate personal letters between the war front and home, Emily and Washington married on January 18, 1865. Four months later at Appomattox, Robert E. Lee, Commanding General of the South (and an engineer by training), surrendered his men to Ulysses Grant, General-in-Chief of the North, and the emergent nation's bloodiest conflicts became history. Little did the newlyweds (who would have one son, John, named after Washington's father) know that they would become the central figures in another of the country's history-shaping events—building one of America's (and the world's) most inspiring engineering marvels.

> "*Probably no greater work throughout history was ever conducted by a man [her husband Washington] who had to work under so many disadvantages.*"
>
> EMILY ROEBLING

For decades it was considered an impossible dream, constructing a bridge across New York's East River (at a point in the river that experienced much of the Atlantic Ocean's severe weather patterns) from Brooklyn to Manhattan. Determined to do it, though, was German-born John Augustus Roebling (1806–1869), a creative engineering genius who had both the imagination and the courage to accept the challenge. Along with his state-of-the-art European engineering education, his impressive portfolio of experience included an extensive list of highly successful, record-setting, long-span bridges. If Mr. John A. Roebling, with all his talent and worldwide reputation, believed such a structure could be built, surely it was possible!

After completing the main plans for the massive bridge shortly after his eldest son's wedding in 1865, Roebling spent the next four years garnering support from leaders in Brooklyn and New York City, and getting authorization from city and state officials in New York, Congress, and the U.S. President (by 1869, that was Ulysses Grant). While all this "politicking" was going on, he sent Washington (his assistant engineer) and Emily to Europe and around the world to learn the latest known techniques about constructing deep-water caisson foundations.

Shortly after their return and prior to the beginning of major construction, tragedy struck. John Roebling, the designer of the Brooklyn Bridge (America's "Great Bridge"), suddenly died, the result of a freak accident at the site of one of the bridge's abutments.

Washington, with Emily at his side, took over as chief engineer to finish all remaining design issues and complete the actual construction of the bridge. Equal to the task, he got construction off to a good start in 1869. But then a handful of unforeseen problems developed, some having to do

The Brooklyn Bridge

An engineering triumph like no other of its time, the structure with a 1,596-foot center span was half again longer than the previous world's record. Its four cables were spun of steel wire, the first use of such in a suspension bridge. Designed by John Roebling, the Brooklyn was completed by his son and daughter-in-law—Washington and Emily.

Photo credit: Evelyn Weingardt

with labor problems, faulty materials suppliers, politics, and public relations. Emily's social skills and quick intelligence were key in untangling many of the snags.

For example, when questions arose concerning Washington's ability to head the Brooklyn project, she successfully defended him by speaking to various groups (including ASCE). Her sophisticated tact, diplomacy, and communications skills won over even the harshest critics (Weigold 1984).

To help her husband even more, Emily began studying (and mastering) civil engineering topics, including mathematics, strength of materials, catenary curves, and cable and bridge construction. Neither Emily nor anyone else could have imagined how necessary this knowledge would become.

In 1872, just three years into the project and during the construction of the Manhattan caisson 24 meters beneath the surface of New York's East River, another devastating catastrophe befell the Roebling family. Washington, a hands-on engineer who spent countless hours below the river's surface alongside his workers, was stricken by caisson disease, a decompression sickness known as the "bends." This disease left him paralyzed, partly blind, and deaf—and regularly unable to speak.

Determined to keep the Brooklyn Bridge—and its construction—in the Roebling name, Washington continued to direct activities from his sickbed. And Emily became his main assistant, the only one allowed to visit with him on a regular basis. As Washington watched the construction from his bedroom window, often through binoculars, Emily made daily inspection visits to the site, conveying messages and information back and forth.

A full 10 years remained to finish construction and, at first, 29-year-old Emily primarily acted as her husband's go-between and field engineer.

As time progressed, however, the number of jobs and tasks she took on increased. Eventually, she was correctly answering so many questions of the bridge workers, suppliers, and engineers—as well as public officials and representatives—that she essentially became the bridge's surrogate chief engineer. So knowledgeable was she about the structure's design and construction that many who watched her work out problems in the field were actually convinced that, indeed, she was the chief engineer.

From 1872 until 1883, when the bridge was finally opened, Emily was the Roebling face the public saw. Her valiant efforts in directing and overseeing the great structure's construction—and working with her bedridden spouse, using her self-taught and on-the-job engineering and construction skills to the fullest—were (and remain) legendary. At the official opening of the bridge in May 1883, New York Congressman (and prominent New York businessman) Abram S. Hewitt proclaimed that Emily deserved equal credit with her husband for being the project's engineer of record.

After the strenuous and demanding ordeal of building the Brooklyn Bridge, which would hold the record (833 meters between towers) for being the longest suspension bridge in the world for years to come, Emily and Washington moved to Troy, New York, for four years until their son, John, graduated from college—Rensselaer Polytechnic Institute, Washington's alma mater. They then moved to Trenton, New Jersey. Although they were quite active socially within a small circle of friends and associates in Trenton, they largely remained out of the spotlight.

Washington Roebling continued his involvement with the Roebling family's cable business. And in time, he became a shrewd and successful investor, highly skilled at dealing in the stock market. Emily, always an outstanding manager, helped make sure family finances were always well cared for and distributed (McCullough 1972).

Additionally, Emily involved herself with diverse personal pursuits, including extensive travel worldwide. In 1896, she was presented to Queen Victoria in London and, subsequently, at court in Russia at the Coronation of Nicholas II. In the late 1890s, during the Spanish-American War, she was one of the leaders and most active members of several national groups and causes, including the American Relief Society.

While her son, John (and his two sons, Siegfried and Paul), remained uppermost on Emily's list of priorities in her post-Brooklyn days, she found time to study law. She received her law degree from New York University in 1899. Over her career, Emily was a much-in-demand speaker and the author of numerous articles on her Brooklyn Bridge experiences and her many other activities.

She also wrote an authoritative biography about her husband, saying, "Probably no greater work throughout history was ever conducted by a man who had to work under so many disadvantages." In praise of his team of engineers (herself included, although Emily refrained from so mentioning her role), she added, "It could never have been accomplished but for the unselfish devotion of his assistant engineers. Each man had a certain department to be in charge of and they worked with all their energies to have the work properly done according to Col. Roebling's plans and wishes, and not to carry out any pet theories of their own or for their own self-glorification" (McCullough 1972).

Emily Warren Roebling, the "silent builder" of the "great bridge," exhibited tremendous grace under fire. She died in 1903 at the age of 60—a full 23 years before her long-ailing husband, who lived until 1926.

Ralph Modjeski

If Ralph Modjeski had become a musician instead of an engineer, the world would not have known one of its finest bridge designers.

Ralph's father was Gustav Zimayer, a theatrical administrator of modest ambitions. His mother, however, was a famous and glamorous international actress. Her stage name was Helena Modrzejewska (Modjeska) and, at birth, Ralph was given her last name and was known as Rudolf Modrzejewski (and, in adulthood, as Ralph Modjeski). In 1876, Helena's career brought her to the United States, where Madame Modjeska quickly captivated the hearts of American theatergoers. She came to be known as "the first lady of the American theater."

Ralph was born on January 27, 1861, in Krakow, Poland, during a time when the country was being fought over by three world powers, Austria, Germany, and Russia. While a youngster in Europe, Ralph studied music under Kazimierz Hofmann, son of the renowned Joseph Hofmann. During his later student days, the teenaged Modjeski associated with Ignacy Jan Paderewski, the illustrious pianist and composer who, as an impassioned patriot, would lead his country to freedom and become prime minister of Poland.

Rather than follow a musical or political path, though, young Modjeski chose a career in engineering, graduating with the class of 1885 from the Ecole des Ponts et Chausses in Paris. Shortly thereafter he moved to the United States seeking employment as a bridge engineer.

On December 28, 1885, Ralph married Felicie Benda at St. Stanislaw's Polish Catholic Church in New York City. They would have three children, Felix, Marylka, and Karolek ("Charles").

Ralph became a naturalized American citizen in 1887, after working two years for the noted bridge engineer George Morison, who some called the "father of bridge building in America." Two enormous steel bridges he worked on while with Morison were the Union Pacific Railroad Bridge over the Missouri River at Omaha, Nebraska, and the Mississippi Bridge near Memphis, Tennessee, which was the largest cantilever bridge in the Western Hemisphere at the time.

In 1893, Modjeski founded his own company in Harrisburg, Pennsylvania. One of the firm's significant early assignments was the development of standard bridge designs for the Northern Pacific Railroad; they ended up being the "gold standard" for many of the railroad's bridges for decades. Another was the firm's commission to design and build a major double-deck railway and highway bridge over the Mississippi River at Rock Island, Illinois (1896). The Rock Island Government Bridge, as it is now known, was a great triumph from inception. It really put Modjeski's company on the move, and Ralph never looked back.

With his company well on its way to success, Modjeski began investing more of his time at professional activities and in holding office in engineering societies. One of the more important society positions he held was the 1903 president of the Western Society of Engineers—at that time, the largest engineering association in the western United States.

In his book *A Man Who Spanned Two Eras*, Gtomb wrote:

Ralph Modjeski

Photo credit: Jan S. Plachta

> He [Ralph] found the time and energy to be active in scientific and professional associations not only in the U.S., but in Canada, Great Britain and France. The list of these groups includes 21 impressive organizations; in some he reached the highest ranks. He was a board member, for many years, in ASCE, the most important engineering organization in the U.S. He was also a board member of the American Institute of Consulting Engineers and New York State Society of Professional Engineers, and one of the founders and a long-time president of the Engineers Club. Additionally, he was active in the American Philosophical Society and Association for the Advancement of Science.
>
> The breadth of his interests is reflected in his social activities. He was never an engineer who existed in the shell of his own specialization. Knowing his family's theatrical history and his artistic abilities, it's not surprising that he was an active member in various cultural groups, including the Chicago Art Institute and Metropolitan Museum of Art in New York, where he was a member of the board (Gtomb 1981).

In the 1920s, Modjeski was chosen chief engineer and chairman of the board of engineers of the Ben Franklin Bridge across the Delaware River, connecting Philadelphia, Pennsylvania, to Camden, New Jersey. At the time of its completion in 1926, this bridge was the longest suspension bridge in the world.

His other best known projects include the Government Bridge spanning the Mississippi River at Rock Island, Illinois; the McKinley Bridge at St. Louis, Missouri; and prominent bridges in Thebes, Illinois, and Memphis, Tennessee. Modjeski and his partners also built the Ambassador Bridge, which links Detroit, Michigan, with Windsor, Ontario, Canada. One of his most famous structures is the Broadway Bridge in Portland, Oregon, spanning the Willamette River. When completed in 1913, it was the country's leading drawbridge and the longest double bascule bridge in the world.

"Ralph Modjeski designed and built over 40 major bridge structures around the country. Most still operate today and are regarded as classic examples of the art of bridge engineering," praised Jan S. Plachta of the U.S. Army Corps of Engineers, Chicago District.

Broadway Bridge, Portland, Oregon

When completed in 1913, the Broadway was the longest double bascule bridge in the world and the last Portland bridge built having trolley rails, which were removed in 1940. Constructed for a mere $1.6 million, the six-span bridge's original deck was made of Port Orford cedar logged from the Oregon coast.

Photo credit: Jan S. Plachta

Modjeski was often called on to consult and advise on projects. After others failed at constructing the cantilever-truss rail bridge in Quebec, Canada, his firm was able to complete it in 1918. Modjeski's engineering feats continued to earn attention and praise. In 1935, as an example, *The Morning Tribune* (in New Orleans, Louisiana), singled out the Huey P. Long Bridge in New Orleans as a "Marvel of Engineering and Work of Artistic Beauty" (MT 1935).

Indeed, Modjeski's skills stood out from others. The best engineers in the country could not build a bridge in the deep delta deposits of the Mississippi River. Because of foundation difficulties, including lowlands, navigation clearances, and strong river currents, constructing this bridge proved to be a pioneering achievement of the first magnitude.

In 1930, the U.S. Deans of Engineering honored Modjeski by naming him one of the 15 top engineers in the world for the quarter century from 1905 to 1930. He was in good company, joining engineers Herbert Hoover, Thomas Edison, and George Washington Goethals, builder of the Panama Canal. Indeed, the press compared the Huey P. Long Bridge to the extraordinary achievement of Goethals.

"No man produced more characteristically American bridges than Ralph Modjeski," wrote David Plowden in *Bridges: The Spans of North America* (Plowden 1974). In fact, his career spanned two eras of bridge design—beginning in the age of steel trusses and grand railways and continuing into the heyday of the suspension bridge and major highways.

In 1931, when he was 70, Ralph and Felicie divorced after many years of separation. He then married Virginia Mary Giblyn, who was several years his junior. The same year, Modjeski was awarded one of his greatest assignments. He was nominated Chairman of the Consulting Board of Engineers for the San Francisco–Oakland Bay Bridge in San Francisco, California. This prestigious position not only represented a tremendous honor, but also clearly reflected the expert builder's high stature in America during the 1930s.

Modjeski's talent and professionalism were constantly acknowledged in the United States and abroad through honorary degrees, medals, and prizes. Technical societies awarded him the Howard N. Potts Gold Medal in 1914, the Franklin Medal in 1922, the John Scott Medal in 1924, the John Fritz Medal in 1930, and the Washington Award in 1931. He received honorary doctorates in engineering from the University of Illinois, from the Pennsylvania Military College, and even from the Polytechnic Institute in Lwow, Poland, his homeland.

Foreign governments acknowledged his excellence. In 1926, the Republic of France decorated him as a Knight of the Legion of Honor. And in 1930, the new Republic of Poland presented him with the Grand Award at the Posen Exposition of Industry and Science. Ralph Modjeski was awarded the coveted Washington Award (medal) by the Western Society of Engineers in 1931 for "his contribution to transportation through superior skill and courage in bridge design and construction."

> "*The big awards presented me, I received quite by accident, and the little ones were given to me because I already had the big ones!*"
>
> RALPH MODJESKI

When once asked about his many awards and honors, Modjeski self-effacingly replied, "Oh, I have the big ones and the little ones. The big ones I received quite by accident, and the little ones were given to me because I already had the big ones!" (Gtomb 1981).

To the end, music played an important part in Modjeski's life. As reported by Gtomb, "Despite deep involvement in his professional work he always found the time to play the piano in the evenings, and he practiced several hours on weekends. As a pianist, he was virtuoso. In addition to Chopin, his favorite composer, his repertoire included the works of Bach, Mozart, Beethoven, Weber, and Schumann. Rachmaninoff was his favorite modern composer."

America's Polish-born, bridge-building icon, Ralph Modjeski died in 1940, but his achievements even now stir and fascinate. The engineering firm he founded (Modjeski and Masters) continues on, still highly regarded for its excellence and innovation.

John Alexander Low Waddell

As one of this country's most honored civil engineers ever, John A. L. Waddell was highly decorated by the governments of China, Italy, Japan, and Russia for his contributions to progress. He also received honorary degrees from several universities: Missouri, Nebraska, McGill, and Puerto Rico—and the Imperial University of Japan. For his contributions to civil engineering theory and practice—in particular, bridge design—he was awarded honorary memberships in the national engineering societies of Spain, Peru, and China, and was a correspondent of the Academy of Sciences, Paris, France, and the Royal Academy of Sciences and Arts, Barcelona, Spain.

Over his career, Waddell coined several popular engineering aphorisms, including, "The science of bridge design lies mainly in the detailing," and "The life of the bridge specialist is by no means easy. His governing motto should be integrity, thoroughness and progress" (H&H 1987).

John Waddell

Photo credit: Hardesty & Hanover, LLP

Born in Port Hope, Ontario, Canada, on January 15, 1854, John grew up in a politically astute family. His father, Robert—a native of Ireland who immigrated to Canada in 1831 at age 16—was appointed high sheriff of the United Counties of Northumberland and Durham in 1865. John's mother was the daughter of Colonel William Jones, the sheriff of New York City and a member of the state legislature.

After studying at a business college in Toronto, John matriculated at Rensselaer Polytechnic Institute (RPI) in Troy, New York. He graduated as a civil engineer, class of 1875, and immediately went to work for two Canadian companies: the Marine Department of Dominion at Ottawa (designing marine structures) and then the Canadian Pacific Railroad.

After moving to the United States a few years later, he first worked in West Virginia and then with Raymond and Campbell, bridge builders in Council Bluffs, Iowa. There began his love for designing large bridges and his fast-rising reputation as a world-class bridge engineer.

From his earliest working days, Waddell wrote technical papers on a wide range of subjects, from railroads and bridges to lighthouses. Because of this and his reputation as a theoretician and educator, having taught for several years at RPI, he attracted the attention of the Japanese government. It offered him the chair of the civil engineering department at the newly founded Imperial University of Tokyo. Prior to sailing for Japan in 1882, John married Ada Everett, the only daughter of Horace Everett, Esq. of Council Bluffs.

During their four years in Japan, John wrote his first book, *The Design of Ordinary Iron Highway Bridges*, which is still used as the seminal text at major engineering schools throughout the world. It would become the "gold standard" for textbooks in the industry.

Returning to the United States, he became the western agent in Kansas City, Missouri, for the Pennsylvania-based Phoenix Bridge Company. While employed by them, he served as a consultant to several municipalities and local railroad companies. One of his designs during this period was the Red Rock railroad bridge over the Colorado River. At 660 feet, it was the longest cantilever span bridge in the United States at the time.

In 1887, at age 33, Waddell founded his own consulting engineering firm in Kansas City. He soon became known for creating daring and unusual structures and, in particular, for movable bridges built throughout America and the rest of the world. His most cited design was the South Halstead Street Vertical Lift Bridge in Chicago, which was completed in 1894—the first important bridge of its kind to be constructed in the United States. After its inception, vertical-lift structures began to be used extensively worldwide.

One year later in 1895, his impressive railroad bridge over the Missouri River, near Omaha, Nebraska, received extensive publicity. With its two 520-foot swing spans, it became the longest bridge of its type in the world. Later, among Waddell's many reinforced concrete bridges and viaducts, the most notable would be his 1913 Arroyo Seco Bridge in Pasadena, California.

In 1907, Waddell was appointed the principal engineer of the Trans-Alaska Siberia Company. This company planned to link European Russia with the United States by building a line across Siberia through a tunnel under the Bering Strait to Alaska and through Canada. The plan was eventually abandoned because of international complications. The dream to link Alaska and Siberia, however, still exists.

South Halstead Street Vertical Lift Bridge, Chicago, 1894

The world's most advanced and modern vertical lift bridge at the time, the 130-foot-long Halstead was the first important vertical lift bridge constructed in the United States. It could be raised to a height of 155 feet above the water. After the Halstead, lift bridges began to be used extensively globally.

Photo credit: Hardesty & Hanover, LLP

By the early part of the twentieth century, Waddell's client list included 50 major railroads and numerous municipalities and governments around the world—and he had designed more than 200 railroad bridges for the Vera Cruz and Pacific Railroad of Mexico and a series of highway bridges in Cuba. His reputation had grown so widespread that industry leaders recognized him as America's "genius in the art of bridge building; one of the outstanding international engineers in his field" (CE 1937).

In 1921, Chinese leaders invited Waddell to judge in a worldwide competition for designing the Peking-Hankow Railroad Bridge over the Yellow River. In 1929, he returned to China as engineering consultant to the ministry of railways and as advisor to the government for civil works.

Waddell was a role model and mentor to many budding engineering superstars, including the great David Steinman (of Mackinac Bridge fame). He conveyed to them that engineers have a moral obligation to share information to improve the profession and society. He wrote voluminously, using long hours on railroad and steamship journeys to record his investigations. His many important books on bridges—in addition to his first one on iron highway bridges—include *Bridge Engineering* (two volumes), *De Pontibus,* and *Economics of Bridge Work.*

> "*The science of bridge design lies mainly in the detailing.*"
>
> JOHN WADDELL

As author of countless state-of-the-art engineering papers, John was awarded ASCE Norman Medals for three of his profession-altering writings in 1909, 1915, and 1918. His 1909 and 1915 award-winning papers delved into the advances in using high-strength steel alloys for bridge design, while his 1918 winner addressed the economics of steel arch bridges.

John's extensive writings dealt with more than engineering. A world-class big game hunter, fisherman, and sportsman, he held several world records and often wrote articles for fishing and hunting publications. Thrice, the Kansas City Whist Club elected him president.

In 1917, Waddell moved his company headquarters to New York City. It was renamed Waddell and Hardesty in 1927 when the brilliant Shortridge Hardesty—hand-picked by John from RPI's stellar class of 1908—became a partner. Currently, the company is called Hardesty and Hanover.

The flamboyant J. A. L. Waddell died in 1938. Six years earlier, at age 78, he had been awarded the Clausen Gold Medal by the American Association of Engineers for his distinguished career and his contributions around the world, and for his advancements to the social and economic well-being of the engineering profession.

David Barnard Steinman

In the middle of the twentieth century, from the mid-1920s to 1960, David B. Steinman was the nation's preeminent American-born designer of the long-span bridges, engineering more than 400 major bridges on five continents—a giant among his peers. He and his archrival, Swiss-born Othmar Ammann (1879–1965), dominated the American bridge-building scene. In their heyday, the two New York–based geniuses designed most of the suspension bridges that were built and had a hand in shaping the majority of the rest of them.

Of the two, Steinman designed a far greater number of major bridges than Ammann. His commissions came from every part of the world, and his designs were regularly hailed for their cutting-edge engineering and design innovations. He refined the use of exposed structural steel as art and pioneered the use of color and illumination on bridges. Over the years, dozens of his structures were honored for being the most beautiful bridges in America and/or the world—many still are.

Although Ammann was responsible for two world record–holder suspension bridges (for longest span between towers), the George Washington Bridge (1931, 1,067 meters) and the Verrazano Narrows Bridge (1964, 1,298 meters), both of which are in New York, Steinman's masterpiece, the Mackinac Straits Bridge in Michigan (1957) held the title for being the world's longest suspension bridge (under-cable) at 2,626 meters. This record stood for more than 40 years, until 1998, when the Akashi Kaikyo Straits Bridge in Japan was completed.

David Steinman

Photo credit: NSPE

The record held by Ammann's George Washington Bridge lasted only six years. It was eclipsed by Joseph Strauss and Charles Ellis's Golden Gate Bridge (1,280 meters, San Francisco) in 1937. Twenty-seven years later, New York retook the title from California—by a mere 18 meters—at the completion of the Verrazano Narrows. The Verrazano (and the United States) lost its world title to the Humber Bridge (1,410 meters) in Britain 17 years later in 1981 (Weingardt 1998).

To Steinman, the Mackinac (the "Big Mack") would always be his crowning achievement. A commemorative U.S. postage stamp of the structure issued shortly after its completion recognized it as an American icon. When Postmaster General Arthur Summerfield personally presented him with a customized, first-issue album of the stamps in a formal ceremony on June 25, 1958, Steinman was in his element.

The unbelievable 1940 collapse of the Tacoma Narrows Bridge ("Galloping Gertie")—designed by Leon Moisseiff (1872–1943), a European-trained engineer who consulted with Strauss on the Golden Gate—sent chills through the U.S. engineering community. Designs of long-span suspension bridges went on hold, and Moisseiff's distinguished reputation was tarnished forever.

The disaster immediately prompted Steinman to study the aerodynamic stability of thin, narrow, long-span bridges. His research resulted in the publication of a series of authoritative articles on the subject and established him as the foremost expert on the aerodynamics of suspension bridges. This expertise, in the long run, is what allowed him to create his masterpiece—the elegant, streamlined, and perfectly aerodynamically stable Mackinac.

A genuine Horatio Alger success story, David Steinman figuratively rose from rags to riches. Born into poverty, he became a financial success beyond anyone's expectations and rose to the highest plateau in his field: one of the greatest bridge builders of all time.

David came into this world on June 11, 1887, a few months after the unveiling of the Statue of Liberty, whose framework was designed by the French engineer Gustave Eiffel. David was the seventh child of Eva Scollard and Louis Steinman, an immigrant laborer family in a blighted tenement neighborhood on New York's East Side. A mathematics prodigy, he was the only family member to attend college. His five brothers and sister followed in their father's footsteps and became factory workers, constantly struggling to make a living.

It was from such abject poverty that the youngest Steinman vowed to escape and make his mark in the world. Early on, he was convinced that getting a comprehensive education, hard work, and thrift were the answers. While a youngster, he, like his siblings and many of his peers, worked at numerous odd jobs for pennies, including selling newspapers under the shadow of the majestic and awesome Brooklyn Bridge.

> *"I would gladly give up all my professional accomplishments to be able to create a single composition of exalting music. If I had my life to live over again, I would correct one omission—I would learn to play a musical instrument."*
>
> DAVID STEINMAN

Completed in 1883, the Brooklyn Bridge, with a clear span between towers of 833 meters, reigned as the world's longest suspension bridge. To young Steinman, it represented the outstanding engineering achievement of his day, and its builders the Roeblings—John, Washington, and Emily—became heroic figures in his eyes. As a feisty, ragged kid, he would mostly get chuckles when he pointed to the Brooklyn Bridge and told those around him, friends and adults alike, "Some day I'm going to build bridges like that!"

While still in high school, David started taking college classes at the City College of New York. There, he completed the first phase of his post–high school education, working at odd jobs to pay for his tuition and living expenses. He graduated summa cum laude with a bachelor of science degree in 1906 at the age of 20.

Immediately, he attended Columbia University and, by obtaining enough fellowships, scholarships, and nighttime jobs to stay the course, he earned three degrees. His education at Columbia culminated in 1911 with a Ph.D. in civil engineering. David's doctoral thesis "The Design of the Henry Hudson Memorial Bridge as a Steel Arch" foretold of an incredible project that would become a reality 25 years later.

In 1910, Steinman accepted a position at the University of Idaho, becoming the youngest civil engineering professor in the country. While there, he published *Suspension Bridges and Cantilevers: Their Economic Proportions and Limiting Spans*. He also translated two German books, *Theory of Arches and Suspension Bridges* and *Plain and Reinforced Concrete Arches*, establishing himself as a prolific academic with a practical bent.

After four years in Idaho, Steinman longed for New York and contacted Gustav Lindenthal, America's leading long-span bridge designer at the time, about working for him on the Hell Gate Bridge, which Lindenthal was in the midst of designing. David was hired on the spot and, on July 1, 1914, became Lindenthal's special assistant, second only to another young bridge-building star, Othmar Ammann. The experience of working together on Hell Gate commenced a rivalry between the two that would last a lifetime.

The guns of war—World War I—started thundering across Europe shortly after David met his bride-to-be, Irene Hoffmann, on a trolley car ride on Long Island. Her father Dr. E. Franz Hoffmann, formerly on the faculty at the School of Medicine in Vienna, approved of Steinman, believing he had good prospects. Plus, he could intelligently discuss Kantian philosophy and world events with him.

Married on June 9, 1915, the young couple would have two sons and a daughter: John, Alberta, and David, Jr. John and David would become physicians specializing in psychiatry, and Alberta would become a renowned psychologist.

After his stint with Lindenthal, which, in addition to Hell Gate, included work on the important Sciotoville Bridge, another well-known U.S. bridge builder, John Waddell, employed Steinman. Waddell's main engineering office was located in Kansas City at the time; David was put in charge of his newly established New York office. While there, Steinman helped design the Marine Parkway Bridge.

From 1917 to 1920, Steinman was a part-time professor of civil and mechanical engineering at the newly formed engineering school at City College. In 1920, he opened his own consulting engineering office. His practice began slowly, and prospects looked quite bleak at the start. Recalled Steinman, "My first fee was $5, and for several months it was a difficult and discouraging struggle. Then Holton Robinson (1863–1945), who built the Manhattan and Williamsburg bridges, asked me to join him in a competition to build the Florianapolis Bridge in Brazil" (Ratigan 1959).

Their design proposal won and they were selected as the project's designers. Thus began a partnership—the firm of Robinson and Steinman—that would, over a 25-year period, design hundreds of impressive bridges around the world before Robinson's death in 1945. The Florianapolis, the largest-span bridge in South America when completed in 1926, was the largest eyebar-cable suspension bridge ever built and the first in the Americas to use rocker towers.

Next for Robinson and Steinman came the Carquinez Strait Bridge northeast of San Francisco, the fourth largest cantilever bridge in the world, and the Mount Hope Bridge over Narragansett Bay, Rhode Island. Commissions for the company quickly started flowing in from everywhere, several from overseas. Neither the 1929 stock market crash nor the Great Depression itself seemed to slow down the newly formed firm.

In late 1929, Steinman and Robinson designed the Grand Mere over the St. Maurice River in Quebec. The project introduced prestressed twisted wire rope-strand cables, a Steinman innovation that later debuted in the United States in 1931, with the simultaneous completion of the pair's St. John Bridge across the Willamette River in Portland, Oregon, and the Waldo-Hancock Bridge across the Penobscot near Bucksport, Maine. The Waldo-Hancock Bridge also featured the first-time use of Vierendeel trusses in bridge towers.

Mackinac Bridge

Stretching 17,913 feet across the Mackinac Straits in Michigan, the Mackinac Bridge was a 70-year dream come true when it was completed in 1957. The $100-million structure's two 24.5-inch-diameter main cables totaling 20,600 miles were spun in 78 days. The "Big Mack" held the title as the longest suspension bridge in the world until 1998.

Photo credit: Franklin Meyers and Gar Hoplamazian

Following those came many other noteworthy bridges such as the Henry Hudson (New York), Deer Isle (Maine), and Thousand Islands (linking Canada and the United States across the St. Lawrence River)—plus a wide assortment of significant structures outside the Western Hemisphere.

In 1947, Steinman was selected to do the reconstruction of the Brooklyn Bridge, the project that had first inspired him to become an engineer. He often said he considered it his supreme accolade to be chosen to modernize the Brooklyn Bridge.

In the late 1950s, Steinman was involved in designing the Messina Bridge, crossing the two-mile-wide Strait of Messina between Sicily and the Italian mainland. It would have been the world's longest suspension bridge by a huge margin. It still remained on the drawing board, however, when, on August 21, 1960, Steinman passed away in his beloved New York City at the age of 73.

A true believer in giving back to one's profession and helping advance it, Steinman served as president of a number of engineering groups, including the New York State Society of Professional Engineers, Society for the History of Technology, and American Association of Engineers (AAE).

As president of AAE, he began a national campaign for more professionalism and stringent educational and ethical standards within the engineering profession—and to get PE registration laws in every state in the union, as well as U.S. territories. He vigorously pushed the concept that engineering was a profession on a par with medicine, law, and science.

In 1934, he invited engineering leaders from four state professional engineering societies—Connecticut, New Jersey, New York, and Pennsylvania—to discuss forming a nationwide society of professional engineers. The result was the formation of the National Society of Professional Engineers (NSPE), for which he worked tirelessly to ensure its success (Robbins 1984).

Specifically, he was its first president (1934–1937, serving two terms), and, in his inaugural or keynote address, he emphasized a need to protect legitimate engineers against competition from the unqualified, from unethical practices, and from inadequate compensation. He sought to build public appreciation and recognition of the engineer.

An inspiring figure on the platform, Steinman made countless speeches on behalf of NSPE and the profession, giving depression-stricken engineers—many without jobs—renewed hope and faith in themselves and their profession. Every engineer could make the profession better than he or she found it by getting involved, he believed. What he promised the nation's engineers was pride in self, pride in profession, and pride in public service (AME 1960).

In addition to being a much-sought-after speaker, David was a prolific and accomplished author, writing both prose and poetry. He was the author of more than 600 professional papers and 20 books, among them *Bridges and Their Builders* (1941) with coauthor Sarah Watson; *The Builders of the Bridge* (1945), a best-selling biography of the Roeblings; and *I Built a Bridge and Other Poems* (1955). His 150-plus published poems included titles such as, "Brooklyn Bridge: Nightfall," "Blueprint," "The Harp," "The Song of the Bridge," and "The Challenge," in which he stated, "Nature said: 'You cannot,' Man replied: 'I can.'"

Over his illustrious career, Steinman received an unbelievable number of prestigious honors and tributes. In the period from 1952 to 1956 alone, he received more than 50 international awards, plaques, citations, and decorations, including the William Procter Prize (American Association for the Advancement of Science) and the 1954 Grand Croix de l'Etoile du Bien (French government). The only other recipient of this award in 1954 was Dr. Albert Schweitzer (Ratigan 1959).

In 1957, he was awarded five major medals:

1. The Kimbrough Gold Medal (American Institute of Steel Construction),
2. The George Goethals Medal (Society of American Military Engineers),
3. The Gzowski Medal (Engineering Institute of Canada),
4. The Louis Levy Medal (Franklin Institute), and
5. The Gold Medal of the Americas (Chamber of Commerce of Latin America).

The first of Steinman's 19 honorary degrees was a doctor of science from his alma mater, the City College, New York, in 1947. His doctor of engineering degree from Manhattan College in 1953 was conferred on him by the most eminent Cardinal Spellman on the occasion of the school's hundredth birthday. In 1957, he received a doctor of law degree from the University of Tampa; at the institution's graduation ceremony, he gave a commencement address titled "Moral Armor for the Atomic Age."

In his later years, Steinman became extremely philanthropic, especially in assisting needy and deserving students by establishing the David Steinman Foundation, the Irene Steinman Scholarship, and the Holton Robinson Scholarship. At City College, the school of engineering building—Steinman Hall—is named in his honor, as are numerous engineering awards programs around the world.

A man with many passions, Steinman was, for one, a skilled horseman, regularly riding his white stallion, Bill, at the head of the University of Idaho's Campus Day parades while a professor there. He shared a stamp-collecting hobby with his youngest son, David; he excelled at photography and loved classical music.

Said Steinman, "When I listen to a composition by Bach, Beethoven, Mozart or Schubert, I would gladly give up all my professional accomplishments to be able to create a single composition of exalting music. If I had my life to live over again, I would correct one omission—I would learn to play a musical instrument" (Raitgan 1959).

In its "turn of the millennium" special issue, *Engineering News-Record* honored Steinman as one of the greatest bridge engineers of all time. And his many outstanding bridges continue to be living monuments to that.

CHAPTER FIVE

Structural Trailblazers

*The first fundamental idea of 'structure as art,'
the discipline of efficiency, is a desire for mini-
mum materials, resulting in less weight, less
cost and less visual mass.*

—David Billington

Of the major branches of the civil engineering profession, structural engineering, over the years, has remained one of the most visible. The
majority of its products—buildings, stadiums, parking garages, towers,
dams, power plants, bridges, and so on—are usually above ground and/or
readily observable by engineers and nonengineers alike.

Along with the wide array of structures civil engineers design, there is
often a tendency to break structural engineering into various subdivisions
when practitioners concentrate on a single type of structure. For instance,
the designers of certain types of structures such as tunnels, dams, or bridges
often prefer the title of tunnel, dam, or bridge engineer. Illustrative of this
are the four engineering luminaries spotlighted in Chapter Four, who were
identified as bridge engineers. This chapter, in contrast, focuses on four civil
engineers who designed all types of structures, predominantly people-use
types of buildings.

Although the media often shortchange structural engineers in favor of
architects when reporting on people-use building projects, the work of such
civil engineers is nonetheless highly visible. At will, these types of engineers can
regularly go to (or drive by) construction sites and admire the fruits of their
labor—and point out noteworthy structural features to all who will listen. It is
one of the extremely satisfying benefits of being a structural engineer.

Next to spectacular bridges, the structures most reported on by the
media are record-setting skyscrapers, sports facilities, space age–looking
buildings, and national monuments—and anything that is the biggest, tallest, longest, or first.

Of the structural-civil engineering giants presented in this chapter, Fazlur Khan and Fred Severud concentrated on building structures and monuments, while Willard Simpson and T. Y. Lin divided their attention between buildings and bridges. All four, though, were (and remain) internationally renowned trailblazers in the overall field of structural engineering and in the design and building of trendsetting structures globally, many of them industry firsts and world record–holders. They not only overcame difficult challenges professionally, but also took on consequential leadership roles in their communities, making them better places to work and live.

Through the ages, structural engineering has always been highly regarded as a noble profession. It is the reason communities have all the awe-inspiring structures that they do—soaring stadiums, towering high-rises, spaceage–looking coliseums and arenas, and state-of-the-art public and private facilities.

Virtually all of the historic seven wonders of the world—from ancient times through the Middle Ages to modern days, from the Egyptian pyramids to the Sears Tower, and beyond—were (and are) structural engineering feats. Without the expertise and talents of structural engineers, the structural performance of those facilities would never have reached their maximum potential.

From two decades after the Civil War—and inspired by structural engineering icons such as Chicago's William Jenney and his first skyscraper in 1885—U.S. structural engineers have been at the forefront of creative, cost-effective, and industry-altering structural designs. Many of the world's leading structural engineers over the years, however, did reside in other countries, and many had a significant influence on U.S. engineers and the development of America's structural industry.

Among the most famous was the French engineer Alexandre Gustave Eiffel (1832–1923), designer of the tower bearing his name and the structural framework that holds up the Statue of Liberty. He was one of the most influential forerunners for using structural iron and then steel in multistory structures in lieu of masonry-bearing walls. He was also a proponent of the merits of theoretical structural analysis. In the early 1870s, Eiffel alleged, "The English engineers have almost entirely bypassed calculations. They fix dimensions of their members by trial and error, and by experiments and using small-scale models. They are ahead of us in this practice, but we have the honor, in France, to surpass them by far in theory and to create methods opening up a sure path to progress, disengaged from all empiricism" (Billington 1983).

Two of the most respected world-class pioneers in the structural design of reinforced concrete structures were Eduardo Torroja (1899–1961) from Spain and Pier Luigi Nervi (1891–1979) from Italy. They were possibly the most aesthetically conscious "structural artist" engineers of their day. Ove Arup (1895–1988), who was born in the United Kingdom to Danish parents, sparked considerable interest everywhere with his structural design solutions for the sail-shaped Sydney Opera House, a complex project that took 16 years to complete (from 1957 to 1973).

In the same time frame, "Mr. Thin Shell Concrete Master-Builder" himself, Spanish-born Felix Candela (1910–1997), excited Western structural engineers with the electrifying concrete shells he was designing and building throughout Mexico in the mid to later half of the twentieth century.

These innovative non-U.S. structural engineers sparked the imaginations of engineers everywhere—but especially in the Americas—inspiring them to regard structural design as an art and to push the engineering design envelope higher and higher.

In addition to the designers of buildings, top U.S. engineers were also greatly influenced by Swiss bridge engineers, in particular, Robert Maillart (1872–1940). Maillart epitomized "structural art"—as defined by Princeton's David Billington in *The Tower and the Bridge*—with the endless new and elegant forms he introduced in reinforced concrete. Among the notable American structural engineering giants whose work was greatly influenced by Maillart was Fazlur Khan. According to Billington, Khan perceived much "similarity between Maillart's ideas and his own" (Billington 1983).

In reviewing the stories of the four American-based structural greats featured here, it is obvious that structural engineering thought and creativity currently have no boundaries. Today's generation of U.S.-based structural engineers have picked up where the non-U.S. engineering community of yesterday left off and pushed the structural engineering envelope forward in great strides. In the future, Americans will increasingly impact structural engineering practices around the world, just as European engineering icons did in the past.

Fazlur Rahman Khan

Considered the father of tubular design for high-rises, Fazlur Khan has become an icon in both architecture and structural engineering. Known best for engineering the Sears Tower in Chicago, he created a legacy of innovative structural designs that is without peer. According to senior engineer John Zils of Skidmore, Owings & Merrill (SOM), "It was his unique ability to bridge the gap between architectural design and structural engineering that truly set Dr. Khan apart from other structural engineers."

Born on April 3, 1929, near Dhaka, Bangladesh (then British India), Fazlur's early resolve to become an engineer was influenced by two people—his father and an older cousin who preceded him into college to study engineering.

Khan's father, a well-known and respected mathematician in British India, taught mathematics and wrote textbooks for high school students. The elder Khan eventually became the director of public instruction in the region of Bengal.

Young Khan received his first engineering diploma from Dhaka Engineering College in 1950. A Fulbright scholarship combined with a Pakistan government scholarship brought him to the United States and the University of Illinois at Urbana. There, he earned two master's degrees followed by a doctorate in structural engineering in 1955.

Khan immediately joined the internationally known architectural and engineering firm of SOM in Chicago. By 1960, he was well into establishing his trademark of creating innovative concepts for tall buildings framed with structural steel, concrete, and/or composite systems.

Fazlur Khan

Photo credit: SOM

His "tube concept," using all the exterior wall perimeter structure of a building to simulate a thin-walled tube, revolutionized tall building design. In 1962, while designing the 38-story, reinforced concrete Brunswick Building in Chicago, he developed methods for interacting shear walls and frames to resist lateral forces. Later, he refined the shear wall–frame interaction system to come up with the "tube-in-tube concept," initially used for the 52-story One Shell Plaza Building in Houston.

Khan's diagonal-framed tube system, first used for the John Hancock Center in Chicago, connected widely spaced exterior columns with diagonals on all four sides of the building. The concept allowed the 1965 Hancock building to reach the then-staggering height of 100 stories. The Hancock Center and Khan's Chicago masterpiece—the 110-story Sears Tower with its bundled tube structure (the world's tallest building for many years)—drew worldwide attention to America's innovations in structural engineering for skyscrapers.

Adding to his reputation are Khan's notable international structures, including the Haj Terminal Building at the Jeddah International Airport in Saudi Arabia—a massive tentlike structure covering nearly one square kilometer of area.

Known for his professional leadership in the field of structural engineering, Fazlur was also active in his local community. For many years, he served on the board of trustees for the condominium development in Chicago, where he lived. And he never forgot his roots.

Khan's homeland came to be called Pakistan in 1947. During 1971, the country was divided into East Pakistan (now Bangladesh) and West Pakistan, with its government and military centralized in West Pakistan. Because of this, the economic condition of East Pakistan (Khan's homeland) deteriorated so much that its people boldly protested the unequal distribution of the country's income and wealth. To discourage unrest, the Pakistani government sent its military into East Pakistan to terrorize the people. Ten million Bangladesh refugees eventually made their way to India while Western countries, including the United States, refused to interfere in "internal politics" (Zils 2000).

As a result, Khan founded a Chicago-based organization, the Bangladesh Emergency Welfare Appeal, to help the people of his homeland. The group, which met on weekends in Khan's home, raised money for aid and circulated brochures to influence the public and Washington officials. Many of the Bengalis involved (including Khan) had family and friends in Bangladesh in obvious danger; Khan's group did what they could to make it safer for them. India's aggressive intervention finally put an end to the killing.

This experience, according to Zils, caused Khan "to become reflective: that all people are citizens of the world and while there are intriguing differences between cultures, people should not be categorized or judged according to these groupings." Khan also believed engineers needed a broader perspective on life, saying, "The technical man must not be lost in his own technology; he must be able to appreciate life, and life is art, drama, music, and most importantly, people" (Ali 2001).

For enjoyment, Fazlur loved singing Rabindranath Tagore's poetic songs in Bengali. He and his wife, Liselotte, who immigrated from Austria, had one daughter, who was born in 1960. A structural engineer like her father, Yasmin Sabina Khan said of her father, "He was always concerned with people and how engineering affected them" (Bey 1998). To keep his memory alive, she wrote a comprehensive book about him and the impact of his work, *Engineering Architecture: The Vision of Fazlur R. Khan*. It was published in 2004, 20-plus years after his death.

Khan died of a heart attack while on a business trip in Jeddah, Saudi Arabia, on March 27, 1982. Only 53, he was a general partner in SOM at the time. He was returned to the United States and buried in his adopted home of Chicago.

Sears Tower, Chicago

At 1454 feet, Sears grabbed the global tall building record from New York City's World Trade Center (1368 feet) in 1974. The 110-story 4.4-million-square-foot building featured a unique bundled-tube structural system to resist both vertical and lateral forces.

Photo credit: SOM-Timothy Hursley

In 1998, the leaders of the city of Chicago, because of their high regard for Khan's accomplishments as an engineer, named a street after him—Fazlur R. Khan Way, a stretch of Franklin between Jackson and Randolph.

It was but one of many recognitions he received during his life and posthumously. For example, in 1999, *Engineering News-Record* honored him as one of the world's top 20 structural engineers of the last 125 years. Three decades earlier, when Khan was 41 years old, the Chicago Junior Chamber of Commerce had named him Chicagoan of the Year in Architecture and Engineering.

Among Khan's other honors were the Wason Medal (1971) and Alfred Lindau Award (1973) from the American Concrete Institute (ACI); the Thomas Middlebrooks Award (1972) and the Ernest Howard Award (1977) from ASCE; the Kimbrough Medal (1973) from the American Institute of Steel Construction; the Oscar Faber Medal (1973) from the Institution of Structural Engineers, London; the AIA Institute Honor for Distinguished Achievement (1983) from the American Institute of Architects; and the John Parmer Award (1987) from Structural Engineers Association of Illinois.

> *"The technical man must not be lost in his own technology; he must be able to appreciate life, and life is art, drama, music, and most importantly, people."*
>
> FAZLUR KHAN

Khan was elected to the National Academy of Engineering in 1973 and received honorary doctorate degrees from Northwestern University in 1973 and Lehigh University in 1980.

Apparent from all of Khan's citations are that his main legacy will be that he, more than any other individual, ushered in a renaissance in skyscraper construction during the second half of the twentieth century. He epitomized both structural engineering achievement and the need for creative collaborative between architect and engineer. To him, for architectural design to reach its highest levels it had to be grounded in structural realities.

Fred N. Severud

One of America's greatest monuments, the towering Gateway Arch in St. Louis, Missouri, stands majestic because of a structural engineer's genius—the same engineer who was responsible for creating the awe-inspiring cable-supported roofs for Madison Square Garden in New York City; Yale Hockey Rink in New Haven, Connecticut; and Raleigh Livestock Arena in Raleigh, North Carolina. This engineer was Fred N. Severud, an intense and brilliant Norwegian-born U.S. immigrant with bold ideas.

Born in Bergen, Norway, into a large family consisting of two brothers and nine sisters on June 8, 1899, Fred was educated at the Institute of Technology in Trondheim, Norway. He moved to the United States shortly after marrying his college sweetheart, Signe Hansen, in 1923. It was in this adopted country that Fred aspired to "fulfill my ambition to become the greatest [structural] engineer in the world" (Watchtower 1982). According to many of his colleagues and, as confirmed by the many lasting, inventive, and structurally historic building structures he designed, it could be said that Severud accomplished his goal.

The structural trendsetter from Norway was more than an innovator of leading-edge structural systems. Anton Tedesko, a distinguished peer and world-class thin-shell expert, said, "He [Severud] was an excellent speaker who inspired complete confidence." Tedesko's remarks were made upon Severud's installation into the prestigious National Academy of Engineering (NAE) in the late 1970s, for his pioneering efforts in the structural field.

Fred Severud

Photo credit: Reprinted from *Engineering News-Record,* copyright The McGraw-Hill Companies, Inc., June 9, 1966

In a letter to NAE (May 19, 1977) accepting the honor, Fred remarked, "Real engineering is like a melody that's softly played in tune. Engineers are often prone to juggle figures rather than ideas. To formulate a closer personal relationship with our architectural clients, an artistic temperament would be a great asset. I suggest that some efforts along this line be included in engineering training." Severud believed that, at the core, "engineers are artists." He regularly asked young engineers who were applying for a job with his company, "What musical instrument do you play?"

"One of Severud's great gifts," said Tedesko, "was that he intuitively understood how structures worked on their deepest level. He was able to clearly see the reasons for structural problems and knew how to prevent them."

Severud was one of the first civil-structural engineers in the world to analyze the forces from, and effects of, atomic bombs. Early in his career, he wrote a comprehensive textbook—*The Bomb, Survival and You*—about the subject and how to provide protection from nuclear explosions. (It was one of several books he would write; all the others would deal with architectural and structural engineering subjects.) His concern about the destructive forces of bombs and explosives partly stemmed from the fact that his brother Bajarne was one of the leaders of the Norwegian underground resistance to the Nazis during World War II (Severud 2000).

The 1965 Gateway Arch in St. Louis

The Gateway Arch honors America's western expansion. The towering 630-foot stainless-steel-faced monument rises from 60-foot deep foundations sunk 30 feet into bedrock. Its double-walled, triangular-shaped legs are stiffened with concrete fill to a 300-foot elevation. Under fierce 150 mph winds, the top deflects less than 18 inches.

Photo credit: Jefferson National Expansion Memorial/National Park Service

Even though Severud was exceptionally intelligent and talented, "he never made others feel stupid," said his son, Fred, Jr., who was also a structural engineer. "Above all else, my father was a kind man. He accepted all as equals and as a result had everyone's highest respect. He had the ability to get the best out of others, not by bullying, but by example and leadership. He was 38 years old when I was born and he died at age 91. In all the time I knew him, I don't recall once seeing him lose his temper. When others let him down or pulled a nasty trick on him, he always could find some excuse for them."

A confirmed atheist in his youth, Severud completely reversed his religious belief system after he arrived in the United States. "His avocation (much more than a hobby) was the Christian ministry," recalled his son. "He spent most of his time outside of working hours preaching and teaching the Bible. He applied his deep religious conviction in every aspect of his life; he didn't just preach it to others, he lived his faith. His love for honesty and justice—and doing the right thing even when it wasn't to his advantage—was passed on to all he encountered."

Severud's decision to become an engineer was not inspired either by his father, Herman, a businessman, or his mother, Cecilia, a homemaker. Originally thinking he would purse a military career, he changed his mind once he started attending the university. There, he decided it would be more exciting to become a structural engineer and design great structures all around the world.

In addition to Fred, Jr., Fred and Signe had three other children: Inger, Laila, and Sonjan.

Severud stayed in touch with many of his university friends throughout his life. Several became as famous as he; Helge Ingstad, for one, became a well-known Arctic explorer and author. And his brother Harald was one of Europe's most important modern music composers. Their accomplishments made Fred extremely proud of his Norwegian roots.

Severud worked for other engineering firms only a short time before he founded his own engineering firm in 1928; called Severud Associates, it was located in New York City. The firm and its founder—in collaboration with well-known architects such as Eero Saarinen (the St. Louis Gateway Arch designer)—created numerous stunning projects through the years.

"I came to America [from Norway] to fulfill my ambition to become the greatest structural engineer in the world."

FRED SEVERUD

In addition to the magnificent stainless steel–faced St. Louis Arch and many cable roof structures, other outstanding Severud-designed projects included two in Canada—Place Ville Marie Center in Montreal and the City Hall Complex in Toronto. Two of his noteworthy early high-rises were Mile High Center in Denver, Colorado, one of the city's first significant post–World War II buildings, and Richard J. Daley Center in Chicago, Illinois, a 32-story civic center structure.

A fellow in the ASCE, Fred received numerous engineering awards such as the Ernest Howard Award and the Franklin P. Brown Medal. The American Institute of Architects (AIA) presented him with its prestigious Honorary Associate Member award for his lifetime of contributions to structural design.

Severud retired from his firm in 1973 just before his seventy-fourth birthday. He spent the remaining 17 years of his life almost entirely dedicated to religious activities. He passed away in Miami, Florida, on June 11, 1990.

Willard Eastman Simpson

Encounters with three great men—two early in his life and one at the beginning of his consulting engineering career—had tremendous influence on young Willard Simpson, both professionally and personally. First was Douglas MacArthur (a Texas high school classmate); next was Massachusetts Institute of Technology (MIT) Professor Charles Swain; and third was Karl Terzaghi, the father of soil engineering.

Simpson was born in San Antonio, Texas, on May 8, 1883, the son of Willard Lloyd Simpson and Edith Carlton. His early education was in the private schools of San Antonio followed by four years at the West Texas Military Academy (later named Texas Military Institute—TMI). There, he learned much about leadership and military-style discipline and was exposed to the decision-making tactics taking root in a young MacArthur (the future five-star general who would oversee the transformation of Japan after World War II). Willard graduated from TMI in 1901 (SAL 1955).

With strong encouragement from his father, 18-year-old Willard applied to and was accepted by the high-status MIT, which he entered in September 1901. Originally enrolled to study naval architecture, Willard quickly switched to civil-structural engineering. One of the main reasons for the switch was that he did not want to stay on the East Coast, where most ship designers were employed. The other was that he was highly impressed by MIT President Henry Pritchett and Professor Swain, both noted civil-structural engineers.

Willard Simpson

Photo credit: Douglas Steadman

Said Willard, "When I was at MIT, the instruction wasn't simply about the teaching of engineering matters, but the whole school was devoted to instilling in us the idea that we would become professional engineers and the leaders in our field. We were taught to continually study and improve our minds—and keep abreast of the times. As professional men [and women] of learning, we were to show the way for others."

According to Simpson, Swain, head of the civil engineering department, would invite his entire class to spend every Tuesday afternoon with him. "He would either read to us from famous books or review the biographies of great engineers and explain how they attained their positions of eminence. He never failed to point out how the most successful of these engineers gave of their time to civic affairs for the benefit of their communities," recalled Willard. "It was many years before I found out how much Dr. Swain's sessions meant to me. I have also learned, through the years, how much it means to have a wonderful garden of friends and that such a garden must be cultivated just as carefully as a beautiful garden of flowers" (Gerhardt 1984).

Simpson graduated with a B.S. in civil engineering with special emphasis in structural design in June 1905. He immediately took a position at MIT as an engineering instructor working for Swain. This additional year in Boston allowed him to absorb more of the city's culture and to cheer on his younger brother, Guy, who was finishing his senior year at MIT.

In 1906, Willard joined the Southern Pacific Railroad in Tucson, Arizona, where, to the skepticism and amazement of other engineers in the office, he introduced two innovations: the contour pen and the Mannheim slide rule. Using the slide rule, he solved railroad curve problems in two days that up until then took others two weeks.

Simpson returned to San Antonio in 1907 to work for architect J. Flood Walker as his structural engineer on the state-of-the-art St. Anthony Hotel. In 1909, Willard and his brother, Guy, established an engineering-construction firm. Among the projects they designed and built were buildings on the new TMI campus at Alamo Heights and at Ft. Crockett in Galveston, Texas. These structures made history by being the first in the nation to be built using concrete tilt-up walls. Soon after, the Simpson brothers dropped construction from their operations and concentrated on consulting engineering, specializing in the design of structures.

In the early days of the twentieth century, structural steel fabricators and concrete reinforcement steel suppliers in many parts of the country often furnished structural design as part of their total bid package. Offering structural design for a fee along with architectural services was a new concept in Texas. At first, the Simpson brothers faced stiff competition from fabricators, suppliers, and other "free" purveyors of structural plans, but they eventually proved that using structural consultants was a better way. As the most outspoken pioneer in advancing this practice, Willard is frequently referred to as the father of the consulting structural engineering profession in Texas.

"We were taught to continually study and improve our minds— and keep abreast of the times. As professional men of learning, we were to show the way for others."

WILLARD SIMPSON

In January 1915, Willard married Mary A. Spencer of Galveston. From their union, three sons were born: Willard, Jr. (who became a structural engineer like his father); Radcliff (who chose an Army career and, after graduating from West Point, served in France, where he was killed in action during World War II); and Bert (who died at age six).

In 1919, Simpson's company branched out in its practice and entered the highway engineering business. The Texas State Highway Department and the U.S. Bureau of Public Roads were just being established, and local governmental agencies needed savvy professional engineers to bring them up to speed on modern road building. In keeping with this, Simpson obtained the commission as county highway engineer for four counties—Kimball, Kerr, Edwards, and Real—in the Hill Country of Texas, northwest of San Antonio.

No contemporary roads previously existed in the four counties, only wagon trails. In helping the area modernize, Simpson engineered hundreds of miles of leading-edge highways, including all bridges spanning the rivers, streams, and gullies along the way. The company's largest bridge at the time was the International Bridge over the Rio Grande River near Laredo, Texas. It—and a similar Simpson-designed bridge over the Santa Catalina River near Monterrey, Mexico—was a high-tech, multispan concrete arch structure never before seen in the area.

In the early 1920s, Simpson pioneered innovative solutions to deal with the unstable, expansive clays* in the San Antonio region. Included among his solutions were machine-drilled, deep foundation piers bearing on solid soils, well below the level of erratic soils.

---

* Expansive clays are those clay soils that expand in volume—and create outward pressures—when saturated. These forces, mostly unwanted, cause structural elements like footings and floor slabs bearing on the soils to heave and move upward and/or sideways, which is often highly destructive and detrimental to the integrity of a structure.

Baylor University Stadium

The stadium, with a capacity of 50,000 when opened in 1950, was renamed Floyd Casey Stadium in honor of the father of Carl Casey who, along with his wife Thelma, contributed $5 million toward the stadium's renovation in 1988. The original post-WWII structure was built mostly with funds raised by the Baylor Stadium Corporation, which was formed to build the football venue. The first game played in the stadium was September 30, 1950, a 34-7 Baylor Bears victory over the Houston Cougars.

Photo credit: Baylor University Athletics

In their report "Drilled Pier Foundations," Woodward, Gardner, and Greer stated, "The earliest record the authors have found of the use of machine-drilled piers for foundations comes from San Antonio, Texas. Willard Simpson, a consulting engineer, recognized early in the 1920s that buildings in that area could be protected from the very destructive effects of soil swelling and shrinkage only if all foundations were supported at levels below the zone of seasonal moisture variations" (Steadman 1999).

Near the end of the 1920s, Simpson apprised the world's great soils engineer Karl Terzaghi of his successful experience dealing with expansive soils. The two became lifelong friends. Simpson's triumphs in successfully designing major structures to deal with poor soil situations were motivation for Simpson to write two landmark papers, "Foundation Experiences with Clay in Texas" (1934) and "Problems in Foundation Design: The Soil Laboratory as an Aid in Their Solution" (1936) (Gerhardt 1984).

During his career, Simpson was responsible for cutting-edge structural design for some of Texas's tallest buildings and most modern structures. Included in his portfolio of history-making projects were Houston's Gulf Building, Austin's Federal Courthouse, El Paso's Natural Gas Company Headquarters, San Antonio's U.S. Post Office, Thomas Jefferson High School, Alamo Stadium and Joe Freeman Coliseum, and Waco's Baylor University (now Floyd Casey) Stadium.

Simpson was president of the Texas section of ASCE and the Texas Society of Professional Engineers. He served on a number of civic boards, including the State Engineer's Committee on Flood Control. He was also on the San Antonio Public Service Board for 11 years, serving 9 years as chairman. His service on the San Antonio YMCA board of directors lasted for 25 years. He was also active with the Boy Scouts, Episcopal Church and Symphony Society of San Antonio—and was a 33rd-degree Mason.

In 1966, when TMI honored Willard with its Distinguished Alumnus Award, its citation acknowledged him as "the engineer who planned and constructed Old Main, which still dominates the TMI campus." He lived his final years across the street from the school's main entrance and was a familiar figure to its cadets, often attending major events and football games. An avid big game hunter and outdoorsman, Simpson remained in fine physical shape until his death on June 7, 1967, at age 84.

Tung-Yen "T. Y." Lin

The name of T. Y. Lin is one of the most recognizable in the world of structural engineering. Called "Mr. Prestressed Concrete," Lin, who died on November 15, 2003, was known globally as one of the greatest creative minds and pioneering leaders in the field—an international authority on long-span construction both in steel and in prestressed concrete.

In 1994, when the California Alumni Association named Lin Alumnus of the Year, Allan Temko, Pulitzer Prize–winning architectural critic for the *San Francisco Chronicle*, reported, "Lin is perhaps the greatest structural engineer in the world, and surely the most fearless" (Temko 1994).

After years of concentrated training and teaching, Lin established his consulting firm in 1954 at age 42. As a consultant, he advanced all known theories and applications—and developed new ones—for using prestressed concrete to build, economically, a multitude of different structures. His firm's prestressed concrete designs show up everywhere. Said Sarah Yang in the *UC Berkley News*, in 2003, "Lin's legacy is international. Almost every continent you go to, there will be structures with T. Y. Lin's mark on them" (Yang 2003).

Starting in the 1950s, Lin believed so strongly in the great applicability of prestressed concrete in construction that he became a relentless force behind furthering its wide acceptance. To that end, he organized the First World Conference on Prestressed Concrete, which was held in San Francisco during the summer of 1957. This very successful conference was attended by more than 1,200 engineers, architects, and contractors from around the world (PCI 2003).

T. Y. told Swan, in 1997, that he considered the development of prestressed concrete to he his greatest engineering achievement. Said Lin, "It was invented by Eugene Freyssinet [1879–1962] and pioneered by the European engineers. They must get the credit for that. But I did write the first readable textbook on the subject, and put all the practice into theory—and later develop the use of it into buildings and bridges" (Swan 1997).

T. Y. Lin receiving the nation's 1986 National Medal of Science from President Ronald Reagan

Photo credit: T. Y. Lin

For this and other achievements, President Ronald Reagan presented Lin with the 1986 National Medal of Science. The award's citation reads: "To T. Y. Lin for his work as an engineer, teacher and author whose scientific analysis, technological innovation, and visionary designs have spanned the gulf not only between science and art, but also between technology and society."

Born in Fuzhou, China, on November 14, 1912, the fourth of eleven children, T. Y.'s first choice was *not* to become an engineer. "I wanted to be a politician!" exclaimed Lin in a 1997 interview in *Bridge Design and Engineering.* "You know, in China to be a politician is to be a statesman, and when I was young, I thought that would be a great way to help the country. But my father, who was a lawyer and a member of the Supreme Court, advised me to stay away from politics." Lin's father instead told him to be become an engineer because that would be a more solid way to help the country and make a living. "The country was backwards [then] and we needed to build many roads and bridges," recalled Lin (Swan 1997).

Lin took his father's advice and enrolled in one of China's leading engineering colleges, Jiaotung University in Tangshan, where he received his bachelor's degree in 1931. Two years later in 1933, he earned his master's degree in civil engineering from the University of California at Berkeley. Returning to China, T. Y. was employed as a design and bridge engineer on the famous Chen-Yu Railroad and Yunnan Burma Highway, eventually becoming its chief engineer. He also worked for the Kung-Sing Engineering Company and Taiwan Sugar Company railroads.

In 1941, T. Y. married Margaret Kao. Her father was also a Supreme Court justice in China. Within five years, the young couple moved to America, immigrating after World War II and becoming naturalized citizens in 1951. They had two children, Paul and Verna. By the time their children were young adults, Margaret had become a tango dancing champion. (Ballroom dancing was a passion shared by T. Y. To ensure that they maintained their footwork skills, the couple had a 1,000-square-foot dance floor installed in their home in El Cerrito, California.)

T. Y. returned to California in 1946 at the urging of his older brother, who was working on his doctoral degree in political science and teaching at Berkeley. T. Y., too, became a member of the faculty at Berkeley, teaching engineering. From 1946 on, T. Y. served continuously on California's faculty, even after he opened his private consulting firm in the 1950s, until 1976. For several years, he was the university's chairman of structural engineering and director of the structural engineering laboratory. At the time of his death, he was professor emeritus of civil engineering at the school.

In a 1984 *RWC Nautilus* interview, Professor Lin had this to say about the narrowness of the education of contemporary civil engineers: "We concentrate so much on mathematics, physics and mechanics, we have no time to teach them real engineering. Four years of college is far too little. It should be more like five to six, maybe even seven years. You can't get the needed exposure to crucial social, legal, financial and environmental issues in a short four-to-five-year program. As an engineer, you are dealing not only with numbers, you are dealing with human beings, so young engineers need to be educated as broadly as possible" (Nautilus 1984).

A great believer that form following function produced beauty, T. Y. said, "Aesthetics is a very important part of engineering. When economics and aesthetics come together—and, if structures fit into their environment—structural beauty results" (Swan 1997).

Over his 60-plus years of experience in consulting, designing, education, and research, T. Y. was the creative force behind numerous award-winning projects around the world, among them the Moscone Convention Center in San Francisco—a massive earth-covered structure having the largest underground room anywhere when built in 1982. Mark Otsea, an architect whose firm, Hellmuth Obata and Kassabaum (HOK), did the architectural design for Moscone, said, "T. Y. was a visionary, always very animated and dynamic, always talking like he was in a hurry and always looking for the simplest, most elegant way to make a space special" (King 2003).

Ten other cutting-edge Lin-engineered structures include

> "*The college training for engineers should be longer, more like five to six, maybe even seven, years. You can't get the needed exposure to crucial social, legal, financial and environmental issues in a short four-to-five-year program.*"
>
> T. Y. LIN

1. Hippodromo National of Caracas Grandstand in Venezuela—a three-inch thick concrete shell cantilevering 90-plus feet,
2. Banco de America Building in Managua, Nicaragua—the country's tallest skyscraper and one of a few buildings that survived the Great Managua Earthquake of 1972,
3. Sky Harbor International Airport Terminal Two in Phoenix, Arizona—giant roof tees create large column-free interiors,
4. Kwan-Du Bridge in Taiwan—a 1,800-foot-long three-span arch structure,
5. Lavern College Campus Center near Los Angeles, California—one of the nation's first noteworthy tensile roof structures,

George R. Moscone Convention Center, San Francisco, California, entrance to underground facility

When completed in 1982, the 275,000-square-foot, column-free, buried exhibition hall was the largest of its kind in the world. Its 16 colossal structural arches, spanning 275 feet, support a roof loaded with three feet of earth and a city park. 5,300 tons of post tensioning—in ties below the floor slab—resist and control the immense horizontal thrust of the arches.

Photo credit: William Andrews, DASSE Design, Inc.

6. UICD Building in Singapore—a 40-story, post-tensioned flat slab structure,

7. Mississippi River Bridge at the Twin Cities in Minnesota—the first prestressed-steel arch bridge in the world,

8. National Convention Center in Jakarta, Indonesia—a steel truss dome with a post-tensioned concrete tension ring,

9. Long Beach Apartments in Long Beach, California—a unique 33-story precast building in the shape of a circle, and

10. Ponce Coliseum in Ponce, Puerto Rico—a massive hyperbolic paraboloid thin shell (4-inch thick) roof structure.

Professor Lin served as a consultant to the U.S. Department of Defense, Federal Housing Agency, California Division of Architecture, Government of Venezuela, Commonwealth of Puerto Rico, and numerous industrial giants such as Lockheed Space and Missile Company, General Dynamics Corporation, and General Electric. A member of the National Academy of Engineering, he also received numerous awards and honors, including the ASCE Opal Award, ACEC Medal of Honor, Freyssinet Medal, and Albert Caquot of France Award.

Lin was the author (or coauthor) of three widely used textbooks: *Design of Prestressed Concrete Structures*, *Design of Steel Structures*, and *Structural Systems for Architects and Engineers*, all of which were translated into several languages. Many of his technical papers are similarly treated as "bibles of the industry." Popular internationally, they have also been printed in several languages. His first major paper—his master's thesis, "Direct Method of Moment Distribution"—was published in 1934 by ASCE.

Never one to be bested, when Lin heard that the great Italian structural engineer Luigi Nervi had built a concrete racing yacht, T. Y. responded by developing a prestressed concrete fishing rod (Temko 1994).

A constantly moving and energized personality, Lin remained a vibrant, creative force in the industry to the end. Until his death, he continued an affiliation with the University of California and was board chairman of Lin Tung-Yen China, Inc., which was headquartered in San Francisco, with branch offices in Beijing and Shanghai.

His final epic project dream to build a bridge—called the "Peace Bridge" by Lin—between Alaska (United States) and Russia still captures the imagination of the public. As T. Y. informed President Reagan in 1986, "We currently have the technology needed—and the engineering skills—to do it today. We just need the will to do it." In discussing his design concepts for the 52-mile-long crossing, Lin said, "Chances are that not much of the structure would be built on location because of the Arctic conditions. But the whole superstructure could easily be prefabricated someplace else using prestressed concrete elements, for instance, and floated into place."

What an exciting concept for the future!

CHAPTER SIX

Daring Innovators

> *A Scientist discovers that which exists. An Engineer creates that which never was.*
> —Theodore von Karman

Progress and advances in civilization go hand in glove with advances in civil engineering. And the profession over the ages has always produced daring innovators when challenged to do so, whether because of nature, humankind, or the circumstances of times in which the engineers lived.

This chapter deals with four engineering greats who exhibited wide-focused vision and the courage to react to such demands. In the process, they enlarged and expanded four different activities of civil engineering and made them major disciplines of the profession. Their efforts brought to the forefront massive rotating and/or dynamic structures, forensic engineering, modern-day seismic design, and computer software and analysis, including three-dimensional imaging.

The first great innovator, George Ferris, undauntedly answered the challenge of the Chicago's 1893 World's Fair Committee for U.S. civil engineers to come up with an answer to the French's splashy Eiffel Tower, the highlight of the 1889 International Exhibition held in Paris. The tower, with its riveted wrought-iron structure left exposed, symbolized the future—and represented a major achievement that dared an equal. By the 1890s, French engineer Alexandre Gustave Eiffel needed no introduction to many in the U.S. engineering community, not only for his Paris tower but also for his unique iron framework that held up the Statue of Liberty.

At a height of 152 feet, Liberty was an enormous artistic and engineering sensation. President Grover Cleveland officially dedicated it on October 28, 1886. But it was not the tallest of existing monuments, not by a long shot. The Great Pyramid in Egypt stood, at 481 feet, as the tallest manmade structure on earth until the 555.5-foot Washington Monument, completed on December 4, 1884, eclipsed it.

America held the title for only five years, though, until the Eiffel Tower, at a height of 984 feet (1,056 feet including the television tower), surpassed the Washington Monument as the tallest structure in the world. The Chicago Fair committee leadership, however, was not interested in a tower, even if it could be taller than the one in Paris. They wanted something different, not an embellished copy of what the Europeans had already done—something representative of America's "can-do" spirit and ingenuity, and its emergence as a world-class nation with vision (Larson 2003).

Civil engineer Ferris gave them just that—an engineering marvel that, hands down, became the "star of the show" of the 1893 Fair. The Ferris wheel, although much shorter lived in the long run, would be even more spectacular than Eiffel's tower had been at the fair in Paris a few years earlier.

Although the Washington Monument held its tallest title for only a few short years, it still remains the tallest masonry monument in the world. South Carolinian Robert Mills (1781–1855), the first professionally educated civil engineer and architect born in the United States, considered winning the 1836 competition to design it one of his greatest triumphs and honors. Quite an acknowledgment considering that Mills, who served as the federal architect and engineer for the U.S. government under seven presidents, was well known for the design of numerous official buildings, among them the U.S. Post Office, Patent Office, and Treasury Building in Washington, D.C.

In addition to Ferris, the three other leading American engineering innovators recognized—Henry Degenkolb, Jack Janney, and Hal Iyengar—also boldly handled the many unknowns presented them in their careers. They unflinchingly rose to the occasion, setting new standards for the profession and providing powerful examples as community role models.

Degenkolb, along with his mentor-turned-colleague-then-partner, John Blume (1909–2002), immensely improved the understanding of earthquakes and the seismic analysis and intuitive thinking needed in the design of large and tall structures. The advancements he made and/or instigated allowed safer construction of buildings and bridges in all the world's highest seismic zones. His activities and publications addressing seismic nuances and design have become the "gold standard" in the field of structural-civil engineering.

After years of research and laboratory work and producing innovative structures, Janney expanded full-time into the investigation, study, and analysis of structural failures and engineering disasters—and brought forensic engineering into the limelight as one of the great branches of civil engineering.

Iyengar incorporated the use of computer analysis into everyday use in the design of tall and complicated structures. Two of Chicago's major landmarks—the John Hancock Building and Sears Tower—were the most recognizable of the many award-winning projects on which he and his SOM colleague Fazlur Khan used leading-edge computer applications and innovations. In 1997, Iyengar's accomplished computer procedures were a key ingredient in the free-form, space age–looking Guggenheim Museum in Bilbao, Spain, being constructed effectively.

George Washington Gale Ferris

George Ferris was a tall, handsome, dapper, and imposing figure. He commanded attention wherever he went long before he became an international figure. As the creator of one of nineteenth century's most imaginative inventions—the Ferris wheel—the young engineer was a legend in his own time.

An 1880 civil engineering graduate of Rensselaer Polytechnic Institute (RPI) in Troy, New York, Ferris worked as an engineer in New York City. He designed bridges, tunnels, and railroad trestles throughout the industrial northeast and midwest before moving to Pittsburgh, Pennsylvania, in 1885. There, the 26-year-old Galesburg, Illinois, native founded G. W. G. Ferris and Company, a steel inspection and civil engineering firm. The following year, he married pretty Margaret Beatty of Canton, Ohio. They would never have children.

Young Ferris was five years old when his father, George Sr., moved the family from Illinois to Nevada—the Wild West—in the summer of 1864, the last year of the Civil War. He grew up on a ranch near Carson City, where, it is rumored, he developed his inspiration for the future Ferris wheel. He was fascinated by—and spent countless hours observing—the large waterwheel at Cradlebaugh Bridge on the Carson River, imagining what it would feel like to ride such a wheel.

George Ferris

Photo credit: Carnegie Library of Pittsburgh

In 1875, George Jr. left home to attend the California Military Academy in Oakland, California. From there, he headed east to attend RPI and fulfill his life's dream of becoming an engineer. While there, his willingness to take on challenges and accept difficult assignments was manifested, both in the classroom and on the sports field. Said RPI Professor Larry Feeser, "Ferris had an admired reputation for invariably winning footraces and being able to throw a ball farther then anyone on campus."

In her *Invention and Technology* article "America's Eiffel Tower," Anne Funderburg described George's personality this way: "He was eminently engaging and social, an entertaining storyteller who often amused his friends with anecdotes. He was an optimist, convinced that he would ultimately overcome any troubles. Even in the darkest times, he was ever looking for the sunshine to come" (Funderburg 1993).

The organizers for the 1893 World's Columbian Exposition (Fair) in Chicago expressed their disappointment that American engineers had not come up with anything "novel and original" to equal the Paris Exposition's Eiffel Tower of 1889. Architect Daniel H. Burnham, head of the fair committee and in charge of selecting its showcase projects, complained at an engineers' banquet in 1891 that, although American architects had come up with great designs, nothing the nation's engineers had proposed would "meet the expectations of the people." Burnham's motto was, "Make no little plans; they have no magic to stir men's minds" (Larson 2003). Within weeks, a personable, confident, well-dressed, 33-year-old engineer with a small engineering firm in Pennsylvania stepped forward—and his was "no little plan."

Motivated by Burnham's taunt, Ferris came up with an enormous revolving wheel, taller than Chicago's tallest building, an awesome magical device that would carry people to breathtaking heights and yet be absolutely safe. One night following an engineering society dinner meeting shortly after the Burnham encounter, said Ferris, "I got out some paper and began sketching it out. I fixed the size, determined the construction, the number of cars we would run, the number of people it would hold, what we would charge, the plan of stopping six times during the first revolution for loading, and then making a complete turn. In short, before the evening was over, I had sketched out almost the entire detail and my plan never varied an item from that day on" (Fincher 1905).

> *"I got out some paper and began sketching it out. I fixed the size, determined the construction, the number of cars we would run ... in short, before the evening was over I had sketched out almost the entire detail and my plan never varied an item from that day on."*
>
> GEORGE FERRIS

At first, people thought Ferris's proposal for a people-carrying, enormous, rotating wheel outrageous—and him to be a wild man—especially when he stated he could not only design but also build the huge contraption in just two years, the time left before the exposition's opening. Some called him "the man with wheels in his head."

Bruce Geno, a Pennsylvania civil engineer and Ferris historian, is quoted in the *Pittsburgh Post-Gazette* as saying it was truly amazing that Ferris got the Ferris wheel "designed and fabricated in such a short time. He used his connections in the steel industry to get steel. Just as impressive, though, was that he was able to convince people it was a good idea to build this monster" (Lowry 2000).

The charismatic Ferris proved that he had not only an inventive mind but also the ability to engineer and build. The Ferris wheel was completed on time and within its $400,000 budget—and it, indeed, proved to be the highlight of the exposition. As the icon of the fair, it became America's answer to the Eiffel Tower.

The Ferris wheel, along with the 1883-opened Brooklyn Bridge, showed that American civil engineering had arrived. American engineers were seen as a force to be reckoned with worldwide.

Ferris had pushed the envelope on how high moving structures could reach. His wheel—which stood 266 feet—was more than 20 stories tall. It had a diameter of 250 feet and a circumference of 825 feet. It was supported by two 140-foot steel towers, which were connected by a 45-foot axle—the largest single piece of forged steel in the world at the time.

Ferris Wheel

The biggest attraction of the 1893 World's Fair in Chicago was the big wheel from Pittsburgh—the 25-story-tall engineering marvel designed and built by George Ferris. In 19 weeks, 1,453,611 customers paid $726,805 to ride in 36 circulating glass-enclosed coaches to a dizzying height of 250 feet above the fairgrounds to view countryside scores of miles away.

Photo credit: Carnegie Library of Pittsburgh

Thirty-six wooden railroad-sized cars—with plush, crushed velvet interiors—held 60 people each. Two 1,000 horsepower reversible engines provided the power for the wheel. Fully loaded, the 1,200-ton Ferris wheel could handle 2,000-plus people and revolve once every 10 minutes.

The spectacular view from its top made the Ferris wheel an enormously exciting attraction. To pay just 50 cents to ride to the "top of the world" was well worth the price. Several weddings and other commemorative events took place on board as the wheel took passengers to dizzying heights.

On the fair's opening day in June 1893, many notables took the first ride, several highly apprehensive at being so far above the ground. They were put at ease soon after the festivities began. At the top of the ride (as reported in the *Pittsburgh Commercial Gazette* in 1893), a "little woman, looking wonderfully pretty in a dainty gown of black trimmed in gold stood on a chair in a car swaying 266 feet above earth, raised a glass of champagne to the others in the car and toasted her husband." In this toast, a beaming Margaret Ferris said, "To the health of my husband and the success of the Ferris wheel" (PCG 1893).

Over the years, Ferris's invention has been replicated often and everywhere. The influence of this engineering and entertainment marvel is exemplified by the countless numbers of Ferris wheels—and other moving entertainment devices of various types and sizes—at fairs, carnivals, and theme parks around the world. The largest Ferris wheel currently is the 443-foot tall, $56.5 million London Millennium Wheel—the London Eye—which began to turn above the Thames River in February 2001.

George Ferris had gained much fame but little fortune with his wheel. And its notoriety, unfortunately, so overshadowed the rest of his engineering accomplishments that Ferris has been remembered as the inventor of only one thing—and not for his many other civil engineering accomplishments.

When reporter Carl Snyder interviewed him for a magazine piece in the early 1890s, Ferris told him, "The firm of which I'm the head looks after or superintends the construction directly of a good number of the steel bridges of the U.S., and it's in the direction of steel bridge construction that my work has been almost exclusively." Stated Snyder, "He [Ferris] greets you easily, his demeanor is quiet, his tones low. For a Western man, he is rather fastidious in his dress. Perhaps his most notable characteristic is his steel blue eyes of remarkable depth and clarity. After listening to his easy, unaffected talk, brilliant without effort for an hour, one feels he is in the presence of a man destined to play an important role in the industrial and mechanical advancement of his country" (Snyder 1893).

Unfortunately Ferris's career was cut short long before his full potential was reached. He died on November 22, 1896, at the young age of 37 of typhoid fever and/or Bright's disease, a kidney ailment.

Jack Raymond Janney

Born and raised in the small town of Alamosa nestled in the mountains of Colorado, Jack was the eldest of three children. His father, who owned and operated a dairy and milk bottling plant in town, was, in Jack's words, "one of the smartest people I've known, even though he only had a sixth grade education."

Janney's decision to become an engineer came about because of his love of mathematics and science. Immediately after graduating from Alamosa High, he enrolled in engineering at the University of Colorado. World War II, in which he served as a Navy pilot, interrupted his education for a few years.

Following the war, Janney returned to Colorado, where he reentered the university and married his high school sweetheart, Margaret "Peg" McKay. The couple would have two sons—Charles (named after Jack's father) and Hugh. Both would become registered professional nurses. While his sons were growing up, Jack was active in their youth groups, especially the Boy Scouts. He also managed little league baseball teams for years, long after his boys became adults.

After completing his bachelor of science degree in architectural engineering, Janney embarked on earning a master's degree in structural engineering with the thought of pursuing a career in the aircraft industry. Those plans, however, changed drastically when Jack's concrete design professor persuaded him to write his master's thesis on prestressed concrete—a relatively new product at the time (1949). As a result, the Portland Cement Association (PCA) hired Janney to head up research on prestressed concrete at its new laboratory in Skokie, Illinois. His work at PCA (from 1949 to 1955) strengthened his interests in research and testing.

In May 1956, Janney formed his own engineering company after consulting on the design of plants manufacturing precast, prestressed girders for 185 bridges on the Illinois Toll Highway System. Since using those girders was a first for Illinois, Janney was also responsible for load testing the project's prototype bridge.

Jack Janney

Photo credit: Jack Janney

One year later, Jack Wiss became a partner in Janney's firm. In 1959, Dick Elstner joined them and the firm took on its current name, Wiss, Janney, Elstner Associates (WJE), headquartered in the Chicago area.

The National Academy of Science (NAS) retained WJE in 1966 to conduct full-scale load tests—some to destruction—of buildings at the site of the 1964 New York World's Fair. Included among the 140 structures was an eye-catching "Unisphere"—a 140-foot, 900,000-pound spherical skeleton depicting earth's seven continents, circled by three giant rings denoting satellites and space travel. "The NAS assignment put WJE on the map," said Janney. It not only established the firm's reputation internationally, but also brought into prominence forensic engineering as a main component of the civil engineering profession.

The firm's range of forensic work and portfolio of investigative, research, and testing projects boggles the mind. Notable projects handled by Janney included the 1981 Kansas City Hyatt Regency walkway failure; 1981 Harbor Cay Condominium collapse (Cocoa Beach); 1980 Las Vegas MGM Hotel fire; 1978 Harford, Connecticut, Civic Center Coliseum collapse; 1979 Rosemont Stadium roof failure (Chicago, Illinois); 1973 Bailey Crossing Apartments collapse (Washington, D.C.); and 1979 Cline Avenue Overpass failure (East Chicago, Indiana).

Over his 50-year career, Janney investigated more than 500 structural collapses and 4,000 cases relating to structural problems all around the world. He also pioneered in the use of three-dimensional structure models as design aids and research tools.

At the height of his career, Janney was known worldwide as "the man who has investigated more structural failures than anyone alive." In the novel *Skyscraper* (Byrne 1984), the hero—a forensic engineer—closely parallels Janney and his career, with two major exceptions. Jack, while a handsome fellow, is completely bald. Plus, he is still happily married to his childhood sweetheart. In contrast, the book's engineer was a carefree bachelor with lots of hair.

A big supporter of alternate dispute resolution procedures, Janney has served as an expert on dispute review boards (DRBs) for many years, resolving construction and engineering problems. His reasoned and pragmatic thinking has helped settle and/or defuse many difficult lawsuits, keeping them from getting out of hand to the advantage of both sides.

Testing structures at the site of the New York World's Fair in 1967

A quiet, thoughtful man with a ready smile, Janney has been a fatherly role model—and patient mentor—for countless young engineers, especially those passionate about leading-edge design and research.

He often relates lessons he has learned from past mishaps in the profession—his own as well as others. He tells of giving his first major speech to a large audience: "I read my talk. It was terrible, a disaster! And my supervisor let me know about it. Never again did I read a speech. From then on, I allowed time for practicing, was always prepared to present and speak fluently."

Two of Janney's biggest pet peeves are the lack of recognition structural engineers are given in the design of building structures and the ever-increasing number of hired gun types of expert witnesses who are eager to testify against fellow engineers.

After the NAS retained WJE to conduct full-scale load tests of the structures at the 646-acre site of the 1964 New York World's Fair, WJE's reputation as an internationally recognized forensic engineering firm was established. Among the Fair's 140 pavilions was a "Unisphere," a 140-foot, 900,000-pound spherical skeleton depicting earth's seven continents, circled by three giant rings denoting satellites and space travel.

Photo credit: Jack Janney

In 1980, Janney retired from WJE, although he remained on its board, and returned to Colorado. He and his wife lived for 24 years in Cherry Hills Village (an exclusive township near Denver), where he was active on its board of appeals and the Charlou Water District. In 2004, Janney suffered a mild stroke and the Janneys moved to Lawrence, Kansas, where their oldest son has his medical practice.

For his stellar accomplishments, Janney received many honors, including the 1985 Distinguished Engineering Alumnus Award from the University of Colorado and the John F. Parmer Award from the Structural Engineers Association of Illinois. He is an Honorary Member of ASCE and, in 1999, was recognized by *Engineering News-Record* as one of the top 20 structural engineers of the last 125 years.

Henry John Degenkolb

At the top of everyone's list of world-renowned seismic authorities is America's Henry J. Degenkolb. When he passed away in 1989 at age 76, his writings about the structural performance of buildings during earthquakes had been published in several languages and were being used as design standards around the world. Some of his papers dealing with seismic design for structures—such as "Earthquake-Resistive Design of Small Buildings" in the 1960 *Proceedings of the Fourth World Conference on Earthquake Engineering*—have become "bibles for design" in the industry.

Henry Degenkolb

Photo credit: Degenkolb Engineers

Degenkolb's devotion to and leadership in improving earthquake-resistive structural design standards and seismic public policy was legendary around the world. He served on countless committees and panels, public and professional, to accomplish this. More than any other engineer, he improved the rational basis for earthquake-resistive design for all types of structures. Stronger collaboration between design engineers and seismologists resulted from his unrelenting efforts.

Born on July 13, 1913, in Peoria, Illinois, Henry was the oldest son of Gustav Degenkolb and Alice Emmert. Although his grandfather was a farmer in Wisconsin and his father was a minister, Henry and his two younger brothers did not follow in their footsteps. All three became engineers instead. Henry said, "I always liked mechanical things. Becoming an engineer was not a conscious thing. I just knew I was going to be an engineer since I was three or four years old."

Although still a student the University of California at Berkeley, the 1933 earthquake in Long Beach, California, captured his attention in a major way. This disaster, coupled with what was learned in the horrendous 1906 San Francisco earthquake, showed the destructive side of nature's forces. So, after graduating in civil engineering in 1936, young Degenkolb itched to engineer solutions that would mitigate damage to life and property caused by earthquakes.

His career began with the San Francisco Bay Exposition Company designing various buildings for the Golden Gate International Exposition of 1939–1940. During this time, two remarkable people came into his life. First and foremost was his wife, Anna Alma Nygren, whom he married on September 9, 1939. They had five children: Virginia, Joan, Marion, Patricia, and Paul.

The other person of great influence was John Gould, chief structural engineer for the Exposition Company—*and* Henry's boss and mentor. After Gould, a highly respected and legendary figure in California engineering circles, founded his own company, he hired Henry as his star employee in

1946. After several years as Gould's chief engineer, Henry became a partner in the firm, which was renamed Gould and Degenkolb. Said Gould, "Even before he became a partner, Henry was the most valuable person in my firm, bar none, including me."

After Gould's death in 1961, Henry continued as president and eventually renamed the firm H. J. Degenkolb Associates, Engineers. Today, it is simply called Degenkolb Engineers. During Degenkolb's leadership, the firm designed numerous multistory structures around the country, many in the San Francisco Bay area. The leading-edge International Building built in 1961, for example, had all the latest, greatest seismic design concepts engineered into it, as did the 1967 Bank of California. Additionally, for the bank's foundation system, Degenkolb built an unusual watertight hull of steel and concrete, three stories deep, to float the building over poor soils (CM 1966).

> *"Engineering, especially earthquake engineering, is a learned profession, as much as medicine, law or theology, or even teaching."*
>
> HENRY DEGENKOLB

What particularly set Degenkolb apart from his peers was his hunger to investigate all notable structures damaged by earthquakes and to find ways to strengthen buildings against future occurrences. Because of his outspokenness about the devastating ramifications of earthquakes, in-depth seismic investigating and analyzing came to the forefront as an important branch of civil-structural engineering design.

Degenkolb constantly emphasized that good engineering judgment was critically important in earthquake design, that simply relying on codes was *not* sufficient to ensure seismic safety. He said, "One central reason is that engineering, especially earthquake engineering, is a learned profession, as much as medicine, law or theology, or even teaching. You could not, for example, write a code of medical practice that gave every detail and would govern everything a doctor does." Often, codes take 10 years to write and a lot of important improvements can occur in that amount of time (Scott 1994).

Degenkolb's first "hands-on" site investigation of the effects of an earthquake occurred in 1952 with the Bakersfield, California, earthquake. Throughout his career, he became involved in analyzing major earthquakes, including Anchorage (1964), Caracas (1967), San Fernando (1971), Managua, Nicaragua (1972), Guatemala (1976), and Mexico City (1985).

Degenkolb was instrumental in founding the Earthquake Engineering Research Institute (EERI) and organizing its First World Conference on Earthquake Engineering. From 1974 to 1977, he served as EERI's president and, over the years, presented several papers at its conferences.

He also contributed to many industry seismic design codes. For example, he was largely responsible for the development of the Applied Technology Council's Document ATC 3-06, "Tentative Provisions for the Development of Seismic Regulations for Buildings" in 1972–1978.

Degenkolb's long-time acquaintances said of Henry: "To most of us, earthquakes are frightening events, but to Henry, they were his laboratory."

International Building,
San Francisco

A classic example of a trendsetting, modern-day, multistory structure designed to resist the devastating seismic forces prevalent in California and other high earthquake areas around the world.

Photo credit: Degenkolb Engineers

Added to Henry's engineering talents were his skills with the camera, allowing him to carefully record details of structural failures.

Said Stanley Scott, who interviewed Degenkolb for two years (from 1984 to 1986) for EERI's Oral History Series, "All of his life, Henry Degenkolb took photographs—photographs of construction, of earthquake damage, of family vacations, of everything. He often had two or more types of camera draped around his neck as he inspected earthquake and construction sites." His photographic archives number more than 30,000 slides, prints, and negatives, documenting major earthquakes between 1936 and 1986 (Scott 1994).

Throughout his career, Degenkolb was involved in leadership roles in a wide array of activities. He was the structural engineering consultant to the University of California, assessing seismic hazards for new and existing buildings. From 1970 to 1977, he was on the Bay Area Conservation and Development Commission and a founding member of the California Seismic Safety Commission.

Degenkolb was also a member of the California State Building Standards Commission (1971–1985) and an advisor to the California Joint Legislative Committee on Seismic Safety (1969–1974). Nationally, he was part of the President's Task Force on Earthquake Hazard Reduction, Office of Science and Technology, in both 1970 and 1978.

From 1973 to 1985, he was the associate editor of the *Bulletin of the Seismological Society of America.*

Degenkolb's professional honors included selection into the National Academy of Engineering and election as Honorary Member in ASCE. In 1967, he was recipient of ASCE's Ernest Howard Award for his preeminence in earthquake engineering.

Srinivasa "Hal" Iyengar

Born into a prominent civil engineering family in Mysore, India, on May 6, 1934, Srinivasa "Hal" Iyengar developed a deep enthusiasm for civil engineering early in life. While still in grammar school, Hal frequently traveled with his father—a civil engineer and chief engineer for the state for 15 years—to various building sites. Responsible for building bridges and dams throughout the state, Hal's father was a great inspiration to him during his formative years.

Srinivasa Iyengar
Photo credit: SOM

In 1955, Iyengar graduated as valedictorian of his class from the University of Mysore, where he received his bachelor's degree in civil engineering. Two years later, he earned a master's degree in hydraulic and civil engineering from the Indian Institute of Science in Bangalore.

By 1957, Hal's interests had shifted toward structural engineering. He moved to the United States to study at the University of Illinois, enrolling in its structural engineering program for a master's degree in the subject. While there, he received a research assistantship sponsored by the U.S. Defense Department. His studies focused on the design and construction of fallout shelters in the event of a nuclear blast.

In 1960, Iyengar met with the chief engineer at Skidmore, Owings & Merrill (SOM) regarding a job. Within 20 minutes, Hal was offered a position with the firm. He stayed there for more than 40 years, rapidly rising through the ranks and eventually reaching the level of partner. Now retired from SOM, he currently serves as a senior consultant in structural engineering.

While at SOM, Iyengar became acquainted with—and worked closely with—the renowned structural engineer Dr. Fazlur Khan, whom he described as "an inspiration and a good friend." Along with Khan, Iyengar helped usher in the widespread usage of modern computers and software as dependable tools in the analysis and design of complicated structures. Their efforts significantly contributed greatly to an industry-wide movement toward the increasing opportunities computers presented for studying greater loading conditions and engineering systems and alternatives.

In the early 1960s, the two SOM colleagues spent much time writing software programs for the firm's first computer, an IBM-1620—a leading piece of equipment in the budding computer industry. Their groundbreaking programs for analyzing buildings changed the way engineers, and eventually architects, created their designs at SOM and in industry. Hal's software programs actually made possible the successful design of Chicago's cutting-edge 100-story John Hancock Center, which was constructed in 1965.

John Hancock Center, Chicago

The noteworthy 100-story, 344-meter-tall Hancock, constructed in 1965, features widely spaced exterior columns connected by huge x-braces on all four sides, resulting in the world's first diagonal-frame tube system. The building's observation deck on the 94th floor allows spectacular views of four states and sights 80 miles away.

Photo credit: SOM-Hedrich Blessing

Iyengar also worked on the 110-story Sears Tower with Dr. Khan, who is generally credited as the main structural designer of this record-setting high-rise (the world's tallest building from 1982 until 1998). Hal considered the Sears Tower his most significant project because, as SOM structural engineer Robert Sinn reported, "The structural engineers were given license to do architecture on a scale with the structure. This project pushed the state of the art at that time, and created a landmark not only in the city of Chicago, but also in the world." Iyengar described the experience as "what it must have felt like to build the Eiffel Tower—to create *art* out of structural technology."

Iyengar also listed the 250,000-square-foot Guggenheim Museum in Bilbao, Spain, as another of his most notable accomplishments. Designed by U.S. architect Frank O. Gehry, the Guggenheim's spectacular structure with its many geometric forms, which required extensive computer time to analyze, dramatically affected the environment around Bilbao. It forever changed the region's landscape. "The building is somewhat spiritual," said Sinn. "It's about awareness, space and architecture working together to create a sense of place that is completely unique," added Iyengar.

Hal, along with other SOM pioneers in the late 1960s, originated the use of structural engineering computer analysis to enhance the architectural aspects of a building, in particular such leading-edge edifices as the John Hancock Center and the Sears Tower. "Engineers must have certain empathy for architecture and be able to visualize the architect's ultimate goal," stated Iyengar. "Artwork is a statement of liberation and an evolutionary trend—much like architecture."

Even though he was at the forefront of the computer age, and deeply involved in the technical preciseness required by the many geometrically complicated structures he engineered, Iyengar spent much time studying art to strengthen his understanding of architecture so he could respond to all architectural styles. He said, "It's imperative for engineers to evolve with today's new technologies and be able to reinvent themselves through the powerful process of computer visualization."

Many of Iyengar's projects are scattered around the world. He frequently revisits them, always looking for economical new engineering solutions for projects on which he currently consults. An in-demand structural peer reviewer, he has done peer reviews for many notable modern structures, including the Petronas Towers in Kuala Lumpur, Malaysia—the world's tallest building from 1998 until 2004.

His involvement professionally includes active participation in both U.S. and international groups—AISC, ASCE (Honorary Member), American Society of Engineers from India, Building Seismic Safety Council, Council on Tall Buildings and Urban Habitat, Institute of Structural Engineers of London (Fellow), International Association for Bridge and Structural Engineering, and Wind Engineering Research Council, Inc.

> "*It's imperative for engineers to evolve with today's new technologies and be able to reinvent themselves through the powerful process of computer visualization.*"
>
> HAL IYENGAR

Iyengar has received numerous awards and accolades over the course of his career, many highlighting his use of computers and other structural advances in forwarding structural design. ASCE honored him with its 1999 Ernest E. Howard Award for his advancement of structural engineering in research, planning, design, and construction. Notable among his lifetime achievement awards is one from the American Institute of Steel Construction (AISC), where AISC credited Iyengar with "having made a science out of steel building analysis."

Engineering News-Record twice named Hal one of its Newsmakers of the Year—first in 1989 for his work on the Broadgate Exchange House, London, United Kingdom, and then in 1997 for his work on the ultramodern Bilbao Museum structural systems.

Hal and his wife, Ruth, have two children, a son and a daughter. His siblings—three older sisters and one younger sister—continue to live in India. Today, Hal spends his leisure time playing tennis and nature hiking, an activity his daughter introduced to him when he was well into his sixties. Among his favorite hiking trips are those to Vermont's Green Mountains and New Hampshire's White Mountains.

CHAPTER SEVEN

Movers and Shakers

> *The credit belongs to the man who is actually in the arena; whose face is marred by dust and sweat and blood; who strives valiantly; who knows the triumph of high achievement; and who at the worst, if he fails, at least fails while daring greatly.*
>
> —Theodore Roosevelt

A good number of civil engineering leaders have, over the years, not only successfully resolved significant issues facing their individual practices and the profession, but also made their mark in society and business. They have reached out, taken on leadership roles beyond the profession, and become citizens of the world. By their example, they have inspired both engineers and nonengineers alike to give back to their profession and their communities, making them better places to work and live.

Although countless civil engineers have been prominently and passionately involved with big-picture issues throughout history, only a few have become national headliners. Most engineers have remained "unsung heroes" for their noble contributions. Even those who have received many honors have been recognized mostly by engineering groups rather than by the public, with their accomplishments featured more in trade publications rather than through mass media outlets.

The Panama Canal, for example, one of the world's greatest manmade wonders ever, could not have been completed as successfully—or in the timely manner it was—without the leadership of civil engineers John Stevens and George Goethals, two giants in the profession. Yet, they are not household names in contemporary society. They should be. So should all civil engineers who have excelled in the uppermost ranks of leadership in industry, corporate America, and the government.

Those civil engineers who were responsible for the success of the moon landing and the interstate highway system—two of the most awe-inspiring engineering triumphs of the twentieth century—should be household names so people can learn more about them and celebrate their deeds. They would certainly be finer role models for America's youth than well-known people who entertain and/or scandalize.

Because passionate involvement in politics is rare among engineers—mainly due to its combative, less-than-perfect, and often unsavory nature—it is time to tip our hats to those civil engineers who have ventured into it. They have made a difference and gotten "marred by dust and sweat and blood" in the process (as articulated by Teddy Roosevelt). Even though the number of civil engineers holding public office is quite tiny, especially considering that engineering is the second largest profession in the country, that handful of civil engineers who are proactively involved have had a major, major influence.

As this is written, there are only two professional engineers (PEs) in U.S. Congress: Joe Barton, a petroleum engineer from Texas, and John Hostetler, a mechanical engineer from Indiana. However, both serve on key committees critical to America's growth and well-being. The country's highest placed civil engineer is currently Andrew Card, Chief of Staff for President George W. Bush. His input and counsel have been especially crucial to the administration's efforts since the terrorist attack on the United States on September 11, 2001.

More than half of the states have one or more civil engineers serving as a state senator or representative, and two-thirds of all state departments of transportation are headed by a civil engineer. Plus, numerous others serve on city councils or county commissions in their localities, and some are mayors of their towns. But, many more engineers held such prominent offices in America's early growth days than today, mainly because of their high level of education compared to the rest of the population and because of their stature in the community and reputations for building awesome projects.

For example, two of Chicago's best mayors in the late 1800s—Roswell Mason (1805–1892) and DeWitt Cregier (1829–1898)—were practicing civil engineers. Notable civil works projects in Chicago were instigated and/or completed when each was the city's highest elected official. In addition to serving as Chicago's mayor, both were active societally and professionally. Mason, an engineer trained on the Erie Canal, was the first president of the Western Society of Engineers (1869–1870) and Cregier was its sixth president (1883–1885). Additionally, Mason was the visionary head of the board of trustees for the University of Illinois for 10 years. Near the end of his term as mayor, he organized the relief and mass rebuilding of the city necessitated by the great fire of October 9, 1871 (WSE 1970).

"The past is the prologue to the future," said Richard Kirby et al. (Kirby 1956). If today's—and tomorrow's—civil engineers are to match the accomplishment of the giants of the past, learning about those who entered the arena that Teddy Roosevelt talked about would be a good place to start.

As Winston Churchill observed, "Those who do not value or appreciate the accomplishments of their predecessors will likely not accomplish anything to be remembered by their successors."

To this end, four outstanding engineers who clearly exemplify twentieth century engineering leaders as inspirational movers and shakers are Kate Gleason, Bucky Fuller, George Clyde, and Al Dorman. Their actions have forever raised the bar for all engineers in connection with societal and industry involvement and leadership. Reading about their accomplishments begs the question, "Which civil engineers on the scene today will raise that bar even higher?" Time, of course, will tell.

Catherine "Kate" Anselm Gleason

Raised in upstate New York in the late 1800s, at a time when women rarely were leaders in industry and commerce, Kate Gleason overcame innumerable obstacles to excel in business and engineering. Among her many notable accomplishments was the invention of mass-produced, low-cost housing built out of concrete. It earned her the nickname "Concrete Kate" as well as a membership in the American Concrete Institute (ACI), the first woman to be so recognized.

Young Kate, although a tomboy who loved riding horses, was a "pretty girl of average height and straight posture with bright blue eyes" (Bartels 1997). She was well read, a wonderful conversationalist, and a witty raconteur—completely at ease in the company of men and women alike. Although she aged gracefully, remaining attractive and vivacious well into her sixties, Gleason never married. She did, however, amass a small fortune—$1.4 million by the time of her death in 1933—during a life that included three illustrious careers: manufacturing, banking, and construction.

Gleason is credited with several noteworthy firsts. She was the first female to enroll in Cornell University's engineering program (1884), plus the first woman to qualify as a member in these eminent organizations:

- Verein Deutscher Ingenieure, the German Engineering Society (1914)
- Rochester Engineering Society (1917)
- American Society of Mechanical Engineers (1918)
- ACI (1919)

Gleason was also one of the first women to be a member of the Rochester Chamber of Commerce (1916) and president of a national bank, the First National Bank of East Rochester (1917–1919). Additionally, she is believed to be the first woman who had an engineering college named in her honor, the Kate Gleason College of Engineering at Rochester Institute of Technology (RIT). During the naming ceremonies in 1998, RIT President Albert Simone remarked, "Kate Gleason was a remarkable woman, an inspiration. She is an example of the determination, hard work and creative spirit that RIT would like to instill in all of our students, male or female" (BPP 1998).

Shortly after naming its engineering college after Gleason, RIT established the Kate Gleason Chair, adding to the 1996 Kate Gleason Scholarship program that provides full tuition for worthy female engineering students. Said Paul Petersen, dean of the college, "The endowed chair and ongoing scholarships will establish a network of Kate Gleason engineering professionals that will provide valuable engineering solutions for the world *and* encourage young women to join the ranks of engineering professionals" (Sheppard 1999).

Kate Gleason

Photo credit: Jan Gleason

Kate's election to membership in ACI resulted primarily from her innovative designs, engineering, and use of concrete in building affordable concrete houses for her "Concrest" and "Marigold Gardens" developments in Rochester—homes she built and sold at a profit for $4,000 each in the early 1920s. The structures employed manufacturing techniques that Gleason learned while studying operations at an automobile factory. She said, "My inspiration for mass production methods came from visits to the Cadillac factory where Mr. Leland showed me the assembly of eight-cylinder engines" (Vare 1987).

To efficiently complete Concrest and Marigold Gardens, Gleason encouraged one of her employees, George Hiller, to develop a lightweight, easily and quickly movable telescoping tower—which he patented—for transporting and discharging concrete.

Gleason also designed, engineered, and built houses (and related structures) in Sausalito, California, and Beaufort, South Carolina. For her post–World War I efforts restoring historic structures in the French village of Septmonts, France, where she owned a seasonal home, Gleason received a major award from the French government—the Croix de Guerre Medal.

Born in Rochester, New York, on November 25, 1865, Kate was the eldest of four children of William Gleason and Ellen McDermot, both of whom were Irish immigrants. Kate's father owned the Gleason Works, a machine tool factory, which he founded shortly after the Civil War. What her father did fascinated Kate and she quickly became captivated by mechanical devices and engineering. She once told an interviewer, "I started reading books on machines and engineering when I was nine" (Bennett 1928).

By the time she had graduated from high school at age 16, Kate had been working (part-time) at her father's plant for four years. Not surprisingly, all of the other workers at his plant were men, so, upon entering college to study engineering, she was not fazed to be the only girl in her class. Her father, reportedly, "had a sympathetic interest in woman's emancipation and an evangelical zeal to acquaint his children—sons and daugh-

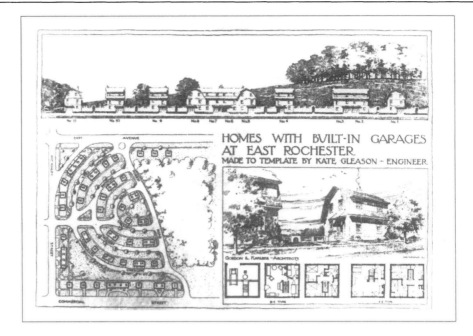

HOMES WITH BUILT-IN GARAGES AT EAST ROCHESTER
MADE TO TEMPLATE BY KATE GLEASON · ENGINEER

Kate Gleason's affordable homes, Rochester, New York

Gleason's innovative design, development, and use of concrete in building affordable homes in the developments "Concrest" and "Marigold Gardens" resulted in her becoming the first woman elected into the ACI, 1919.

Photo credit: Jan Gleason

ters alike—to the marvels of engineering." Among her mother's friends were women like the suffragist Susan B. Anthony, who influenced many concerning the role of women in society.

About Anthony, the competitive Kate said, "She impressed one fact upon me while I was growing up: Any advertising is good. Get praise if possible, blame if you have to. But *never stop being talked about*" (Chappell 1920).

William Gleason, although a master machinist, mechanic, and inventor (he invented the first bevel gear planer) was not exceptionally gifted as a businessman or salesman. Before his oldest daughter could finish college, he called her home to help with his struggling company. Kate convinced him to concentrate on his bevel gear planer, which produced beveled gears faster and cheaper than any other methods—something the fledgling auto industry, in particular, needed immensely. At this point, she became the superstar of Gleason Works—its traveling salesperson and best financial mind. Some called her the "Madame Curie of machine tools" (Colvin 1947).

She put the company on the map internationally by expanding into Europe in the early 1900s. This success came because Gleason introduced a planer product that was better engineered than the competition's. Called Gleason Corporation today, the firm continues to operate manufacturing plants in England, Germany, and India as well as America. Two-thirds of its total sales come from outside the United States.

In 1914, Kate left Gleason Works because she was appointed by the bankruptcy court of New York to serve as receiver—the first female to ever be so named—in the Ingle Machine Co. case. Before she took over and reorganized the failing company, it was $140,000 in debt. When Kate returned the company to its shareholders three years later, it had earned $1 million. Shortly after that, she was elected president of the First National Bank of East Rochester and her second career was well underway. It eventually led to her third career as a builder and developer.

In a 1926 interview, this renowned 60-year-old industrialist and engineer stated that being a woman was never a disadvantage in her chosen field: "When I recall stories told me by women struggling for a place in other professions, I insist that engineers are in a class apart." Gleason said that, in her profession, ideas were objectively judged. A good idea is a good idea, no matter who comes up with it—man or woman (Bennett 1928).

One of Kate's last forays into international engineering came when, at age 65, she served as the ASME representative to the World Power Conference in Germany in 1930. Gleason died peacefully on January 9, 1933, in Rochester, her beloved hometown.

> "*When I recall stories told me by women struggling for a place in other professions, I insist that engineers are in a class apart.*"
>
> KATE GLEASON

Richard Buckminster "Bucky" Fuller

Back in 1959, the distinguished guest of honor—although standing less than five and a half feet tall with a shiny bald head and patches of white hair—had a commanding, electrifying presence. Before he began his talk, he said, "Just call me Bucky." The standing room only audience at the University of Colorado was completely enthralled. Although Fuller began his hypnotic, thinking-out-loud talk on art, humankind, engineering, and the environment with little fanfare, he immediately became a moving dynamo on stage. He captivated listeners with his nonstop gestures and rapid-fire delivery. Many of his words and terms—like synergetics, Dymaxion charts, and Spaceship Earth—had never been heard before.

Fuller came to Colorado at the request of Professor Brown, head of the university's Fine Arts Department and one of Bucky's long-time friends. Members of Dr. Brown's seventeenth-century art history class (which included the author, a civil engineering student taking the course to fulfill a humanities requirement), were special guests at the great event. This amazing presentation by an amazing man—an engineer, mathematician, inventor, architect, philosopher, global thinker, visionary, poet, and cosmologist—was unforgettable.

Labeled a modern-day Leonardo da Vinci and the Thomas Edison of his times, Bucky was one of the key innovators of the twentieth century. He was also controversial, a free spirit who became a favorite of the restless youth of the late 1960s and 1970s. Fuller was one of society's first futurists. His work paved the way for many who followed (Alvin Toffler, Joel Barker, etc.).

Although best known universally as the inventor or engineer who made the geodesic dome popular, Fuller's influence was more far-reaching. Much of his work dealt with exploring and creating synergy,* which he found to be a basic principle of all interactive systems. As the developer of the subject he called synergetics—a "geometry of thinking"—he was the person most responsible for popularizing the terms "synergy" and "ecology."

Bucky Fuller looking through a "tensegrity" model

Photo credit: Boston Public Library, Print Department

Fuller created what he called "comprehensive anticipatory design science," which anticipates and solves humanity's problems by providing more and more life support for everyone, with fewer and fewer resources. He popularized the phrase "doing more with less." He routinely demonstrated his ideas in "artifacts," tangible prototypes or models of designs and principles.

His Dymaxion† map of the world (featured in *Life* magazine in 1943) was the first in the history of cartography to reveal the whole surface of the earth in a single view. It showed the continents on a flat surface without visible distortion—with earth as one island in one ocean.

Driven by the belief that humanity's major problems were hunger and homelessness, Fuller was committed to solving them through inexpensive and efficient design. He was optimistic that, through research and development, responsible engineering, and increased industrialization, humankind could generate wealth so rapidly that all people could live in peace and prosperity. He said, "Making the world's available resources serve 100% of an exploding population can only be accomplished by a boldly accelerated design revolution" (Sieden 1989).

Fuller was one of the earliest proponents of renewable energy sources, including solar, wind, and wave, which he incorporated into many of his designs. He claimed, "There is no energy crises, only a crisis of ignorance." His research demonstrated humanity could satisfy 100 percent of its energy needs while phasing out fossil fuels and atomic energy. In the 1970s, for instance, he showed that a wind generator fitted to every high-voltage transmission tower in the United States would generate three and a half times the country's total power requirement.

A sixth-generation New Englander and grandnephew of American transcendentalist Margaret Fuller, R. Buckminster Fuller was born on July

*Synergy: Arrangements that are mutually beneficial to the parties involved and/or a combined entity that has a value greater than the sum of the parts (or the whole is more than the sum of its parts). In Fuller's words, "Synergy means the behavior of whole systems unpredicted by the behavior of their parts taken separately." His favorite illustration was the behavior of alloys (like chrome-nickel steel): "Synergy alone explains metals increasing their strength."

†Dymaxion: A trademark Fuller word, the use of which was first widely circulated by Marshall Fields department store, in the late 1920s, to promote an exhibit featuring Fuller's revolutionary spherical house design. The word was frequently used by Fuller to mean "doing more with less."

12, 1895, in Milton, Massachusetts, the son of Richard B. Fuller and Caroline W. Andrews. Bucky graduated from Milton Academy High School, attended Harvard University and the U.S. Naval Academy, and served in the U.S. Navy in World War I.

In 1917, on his twenty-second birthday, he married Anne Hewlett on Long Island, New York. They would have two children, Alexandra and Allegra, and would live in many places because of his varied jobs and whims. Shortly after World War I, Bucky went into the construction business with Anne's father. In less than 10 years, poor economic conditions forced him out of the company.

Despondent over that and other business failures—and grieving because one of his young daughters had died unexpectedly—Fuller came to a defining moment in his life. Standing on the banks of icy Lake Michigan in 1927, he resolved to dedicate his energies to the search for socially responsible answers to the world's problems. Thus began a 56-year experiment to discover what an unknown, penniless individual could do for humanity through effective engineering and design.

His pursuit to develop answers to global problems doing more with less was based on the design principles of nature. It made Fuller the pioneer of whole systems thinking, analysis, and design, which caused him to refer to himself as a "comprehensive anticipator"—a "design scientist" for "Spaceship Earth."

Fuller, who was the inventor of the Dymaxion house and car and the recipient of 26 U.S. patents, even created a popular-in-his-day board game—the intriguing "World Game," which used his Dymaxion map displaying world resources and populations. The game lets players strategize solutions to world problems, matching human needs and trends with the earth's assets.

Among Fuller's most famous books—he wrote 28 books and thousands of articles—are *No More Secondhand God* (1963), *Ideas and Integrities* (1963), *Operating Manual for the Spaceship Earth* (1969), and *Earth, Inc.* (1973) in which he wrote, "In reality, the Sun, the Earth, and the Moon are nothing else than a most fantastically well-designed and space-programmed team of vehicles. All of us are, always have been and so long as we exist, always will be—nothing else but—astronauts." In his Spaceship Earth book, Fuller wrote, "Man must be educated into realizing his tremendous potential as a universe-exploring being" (Fuller 1969).

In 1946, after his Dymaxion automobile and house ventures, Fuller accepted a position as a professor at a small but progressive college in North Carolina—Black Mountain College. There, he instigated a revolutionary structural design that would make him famous internationally and advance the fields of engineering and architecture. In 1949, he erected the first geodesic dome building in the United States. Comprised of a series of tetrahedrons (triangular pyramid shapes), the 14-foot diameter dome was constructed using lightweight aluminum aircraft tubing with a vinyl-plastic skin. His geodesic dome revolutionized the industry's thinking about the efficiency of structures.

To prove the soundness of his design, Fuller and several students who had helped build it would hang daringly from the structure's framework to awe his nonbelievers.

Fuller's geodesic domes are a complex assemblage of triangles in which all structural members contribute equally to the whole, form a spherical shape, and grow stronger as they grow larger. They can sustain their own weight with no practical limits and have the highest ratio of enclosed area to external surface area. When complete, the structures—especially very large ones—weigh less than their parts because of the air mass inside the dome. When it is heated warmer than the outside air, it has a net lifting effect like a hot-air balloon.

> "*Man must be educated into realizing his tremendous potential as a universe-exploring being.*"
>
> BUCKY FULLER

In his later years, in conversations with structural engineers and architects about the their efficiency of their buildings, Bucky would always ask the question, "What does your structure weigh?" If they did not know or if it was heavy, he would chide them for wasting materials.

In 1953, Fuller and his patented geodesic dome were elevated to international prominence when the first conspicuous commercial geodesic dome was constructed. A 90-foot diameter hemisphere enclosed the courtyard of the Ford Motor office building in Dearborn, Michigan, so the area could be used all year-round. The originally designed structure weighing 160 tons was scrapped because it was too heavy to be supported by the existing building walls and foundations. Fuller's dome, weighing 8.5 tons, became the solution. Plus, it was erected within weeks, so it could be used for the Ford Motor Company's fiftieth-anniversary celebrations. Media from around the world gathered to report on Ford's anniversary activities. Of course, they marveled at the dome's construction, and word of Fuller's deed quickly spread.

The U.S. government also helped broadcast Fuller's work. Recognizing the practicality of his structure, it contracted with him to build small domes for the armed forces. The U.S. Marine Corps hailed the geodesic dome as the "first basic improvement in mobile military shelter in 2,600 years" (Sieden 1989).

Within a few years, Fuller's domes were showing up everywhere. In 1957, a large geodesic dome for an auditorium in Honolulu, Hawaii, was put up so quickly that, 22 hours after its parts were delivered, a full-house audience was comfortably seated inside the building enjoying a concert.

Today, more than 500,000 geodesic domes (and variations of them) of all types and sizes dot the globe. Notable ones include the 265-foot-wide Epcot Center at Disney World in Florida; a 360-foot-tall dome over a shopping center in downtown Ankara, Turkey; and a 280-foot-high dome enclosing a civic center Stockholm, Sweden. For many years, the world's largest aluminum dome located in Long Beach Harbor, California, housed Howard Hughes's massive airplane—the Hughes Hercules, which was rather derogatorily dubbed the "Spruce Goose" by many.

Innumerable plastic and fiberglass "radome" weather stations enclose delicate radar and sensitive equipment in the Arctic, Antarctic, and cold regions around the world, withstanding extreme cold temperatures and winds up to 180 miles per hour. Corrugated metal domes provide low-cost shelter to families in third world countries. Domelike sculptures grace many civic centers, parks, and playgrounds.

Possibly the most famous geodesic dome of all time was the 20-story dome housing the U.S. Pavilion at Montreal's Expo '67. Shortly after that success, Bucky presented the feasibility of a geodesic dome two miles in diameter that would enclose midtown Manhattan in a temperature-controlled environment. It would pay for itself within 10 years from the savings of snow-removal costs alone.

In 1959, Bucky was appointed research professor at Southern Illinois University. Almost 10 years later, in 1968, the board of trustees appointed him university professor, only the second person in the school's history to be so honored.

During his career, Fuller received 47 honorary doctorate degrees and numerous awards, including the U.S. Medal of Freedom, National Institute of Arts and Letters' Gold Medal Award, AIA Gold Medal, and Royal Gold Medal for Architecture awarded by Her Majesty the Queen of England. In 1969, Fuller was nominated for the Nobel Peace Prize. In 1999, when *Engineering News-Record* named Fuller among its 20 greatest structural engineers of the last 125 years, it described him as "an engineer and prophet who believed that technology could 'save the world.'"

Spaceship Earth at the Epcot Center, Orlando, Florida, 1982

From day one, Fuller's 160-foot geodesic sphere served as the Center's theme structure. Considered one of the most efficient structural systems ever devised, it has been hailed as the hope for humanity's future housing—and other structural needs—while being responsive to environmental and sustainable development concerns.

Photo credit: Frank Heger of Simpson, Gumpertz & Heger, Inc., structural engineer of record for the project.

Content with the impact his geodesic domes were experiencing worldwide and having proved that even his most controversial ideas were practical and workable, Bucky spent the final 15 years of his life traveling and lecturing on ways to better use the world's resources. He presented workshops to millions of people, lectured at 550 universities, and, in the course of his work, circled the globe more than 50 times. He died on July 1, 1983, at age 87.

Fuller was a man ahead of his time, viewed by some as an impractical dreamer and embraced by others as a visionary genius and a brilliant engineer. According to the Buckminster Fuller Institute, even Albert Einstein was prompted to say to him, "Young man, you amaze me!"

Although Fuller maintained he was a "comprehensivist" interested in almost everything, his life and work were dominated by shelter, housing, and transportation. In the final analysis, Fuller was a dogged individualist who profoundly affected the awareness of the emerging social and environmental potential of humanity—and whose engineering genius continues to be appreciated worldwide.

George Dewey Clyde

A take-charge, no-nonsense, two-term governor of the state of Utah (1957–1965), George Clyde was the first practicing civil engineer—and registered professional engineer (PE)—to hold such a lofty public position during the twentieth century. Long before President John F. Kennedy made his "Ask not what your country can do for you, ask what you can do for your country" speech, Clyde was immersed in doing just that, locally, statewide, and nationally.

Prior to his election to Utah's highest office, Clyde was much sought after to fill public leadership roles in several areas, in large part because of his expertise in water development and conservation and because of his reputation as one of America's preeminent irrigation and hydraulic engineers. In 1945, for example, he was named chief of the Division of Irrigation Engineering and Water Conservation and Research for the U.S. Soil Conservation Service in Washington, D.C., and in 1953, as head of the Utah Water and Power Board.

While serving in those highly visible positions of power, Clyde was responsible for instigating the planning, design, and construction of several notable water projects in the West. He was one of major forces behind the building of two crucial U.S. Bureau of Reclamation (USBR) dams—the Flaming Gorge (Utah, 1964) and Glen Canyon (Arizona, 1963).

While governor, Clyde modernized Utah's state highway system and revamped its state road commission, creating a more efficient and professional department. He also broke ground for President Dwight Eisenhower's federal interstate highway system through Utah. During his term, state highway construction increased by 500 percent.

In addition to road and bridge projects, Clyde secured funding for the design and construction of numerous needed public facilities. He was a major factor in the building of the University of Utah's world-renowned medical school and in establishing Utah as the trendsetter among western

states for public libraries, schools, and park systems, including Canyon-
lands National Park.

George Dewey Clyde was born in Mapleton, Utah, on July 21, 1898, to
Elanora Jane Johnson and Hyrum Smith Clyde, a farmer. Active Mormons,
Elanora and Hyrum instilled a strong work ethic and community spirit in
their children, especially George, who exhibited a passionate interest in scientific analysis early on.

After graduating from
Springville High School, he
attended Utah State University (then called Utah State
Agricultural College), receiving a bachelor's degree in agricultural engineering. He then
earned a master's degree in civil
engineering from the University
of California at Berkeley.

In 1919, he married Ora
Packard. They would have
five children, one of whom,
Ned, would later cofound
Woodward-Clyde, which would develop into one of
the world's largest and most respected geotechnical-
environmental engineering firms with numerous offices
around the world.

*Governor George Clyde
signing legislation officially
renaming his alma mater,
Utah State University, 1957*

Photo credit: Special Collections & Archives,
Merrill Library, Utah State University

Armed with both a master's degree and a bachelor's
degree in engineering, George commenced his engineering career in 1923 by joining the Utah State engineering
faculty in Logan, Utah. There, he would teach continuously until he was
called into full-time public service. After several years as a professor at the
institution—specializing in agricultural, irrigation, and hydraulic and water
engineering—Clyde was named dean of engineering, a post he held until
1953. His advice was often sought by a wide range of community leaders
on matters dealing with his expertise.

In 1934, Utah Governor Henry Blood called on the 36-year-old civil engineer to fill the position of state water conservator to deal with Utah's worst
drought in contemporary times. Even though the Great Plains states—Utah
among them—were especially hard hit by the 1934 drought, its impact was
felt nationwide. Along with the drought came a series of seemingly endless
windstorms creating the terrible post-Depression Dust Bowl years.

The Dust Bowl era dramatically increased suffering among Americans
nationwide, especially those already left destitute by the Great Depression.
Banks closed everywhere, and countless farmers lost hope, abandoned their
lands, and moved on, some to the West Coast, others to beleaguered cities—all desperately in search of a better life.

In Utah itself, from June 1933 to May 1934, rainfall was barely 50 percent of normal. As state water conservator, Clyde's analysis of the state's water

Glen Canyon Bridge and Dam under construction

The undertaking produced Lake Powell, Arizona-Utah, 1964, an integral part of the massive Upper Colorado River project, built to cope with the arid west and control the Colorado.

Photo credit: Richard Weingardt

prospects in 1934 revealed that the state's lakes were at historic lows, with the supply of irrigation water around a third of normal (Reeve 1995).

Clyde calculated that, while on average, around four million acre-feet of water ran through Utah's canals each year, less than one million acre-feet would be available in the future. He further revealed that few crops would be harvestable, and only one cycle of alfalfa, a much-needed crop for the state's livestock industry, would likely mature. The state's sheep and cattle herds, so important to the area's economic well-being, faced an industry-threatening situation—and the state, as a whole, faced a strong likelihood of a mass migration.

Clyde's report, along with his solutions on how to resolve the problems, were forwarded to Washington, D.C., post haste by Blood and Robert Hinckley, Utah's director of the Federal Emergency Relief Administration (FERA). Within 36 hours, President Franklin Roosevelt approved an emergency grant of $600,000 for the state. These funds were quickly depleted and, following another appeal to FERA officials, the state received an additional $400,000 (Fuller 1992).

In a little over three months, $1 million in federal assistance was distributed to a myriad of crucial projects identified by Clyde, Blood, and Hinckley. Nearly 300 wells were sunk, 120 springs developed, 185 miles of irrigation ditches lined, and 100 miles of pipeline laid. Plus, long-term solutions to drought—current and future—were initiated, including the construction of dams and numerous reclamation projects statewide.

Soon after Clyde addressed Utah's drought emergency, he was appointed to the advisory board of the Utah Department of Industrial Development Water Resource Division. Shortly after, he was elected director and vice president of the Utah Water Users Association.

Even before construction of the massive, history-making Hoover Dam began in the early 1930s, Clyde was a strong supporter of controlling and using the waters of the powerful Colorado River before they discharged into the Gulf of California and the Pacific Ocean, especially the section of it running along the borders of Utah and Arizona. A key element in this grand plan was the proposed Upper Colorado River Storage Project. In an effort to finally obtain federal approval for the project, the powers that be in Utah decided that Clyde—with his national reputation for resolving important water issues—was needed as head of the Utah Water and Power Board.

After taking over the position in 1953, Clyde—with the help of U.S. Senator Arthur Watkins (R-Utah), a close friend of President Eisenhower—convinced the administration and U.S. Congress of the immense value of the Upper Colorado venture. Legislation to build it was officially signed into law on April 11, 1956, and final design and construction began immediately on both Flaming Gorge and Glen Canyon.

In the process, Clyde became well acquainted with the talents of USBR's chief designing engineer, the legendary dam expert John L. Savage (1879–1967). Most notable of the dams the internationally renowned Savage was involved with were the Hoover and Grand Coulee. After a visit to the Yangtze River Gorge in the 1960s, Savage proposed a layout for the project now called Three Gorges Dam. Currently near completion, it is the largest dam ever built.

As teacher, engineer, and politician, Clyde wrote on a wide range of subjects, including his oft-quoted article "History of Irrigation in Utah" (1959, *Utah Historical Quarterly*), and was the recipient of numerous public, industry, and engineering honors. In 1962, midway through his second term as governor, he was presented with an honorary doctor of law degree from the University of Utah, an accolade he, as a pragmatic engineer, took great pleasure in.

> *"The professional success of engineers is measured by what they ultimately accomplish in engineering, how honest and ethical they are, and how much they contribute to their profession and society."*
> GEORGE CLYDE

Clyde declined to run for a third term in 1965, and calmly, without fanfare, left Utah's highest office and public life forever to once again become a private citizen. He joined his son Ned's burgeoning engineering company and spent the rest of his career as an engineering consultant.

Frank Waller, the last chairman of the board for Woodward-Clyde before it was absorbed by the engineering giant URS, recalled having dinner with Ned and the former governor in 1970:

> Ned was a great believer in taking care of business, but not at the expense of your family, and you could see where he got that philosophy. His father [George], a true gentleman and extremely alert at 72, believed, 'The professional success of engineers is measured by what they ultimately accomplish in engineering, how honest and ethical they are, and how much they contribute to their profession and society. But that success ends up being pretty hollow if a person's family life has to suffer in the process.'

On April 2, 1972, the 74-year-old professor-governor-engineer quietly passed away at his home, after having suffered a debilitating stroke one year earlier. However, the tremendous impact of his leadership, both within and beyond engineering, will last for all time.

Albert "Al" A. Dorman

A man of broad vision and many accomplishments both within and outside the field of civil engineering, Al Dorman is the only person ever to be both an Honorary Member of ASCE and a Fellow in AIA at the same time. During his distinguished 50-plus-year-long career, he worked on all seven continents and had a college named in his honor—the Albert Dorman Honors College at the New Jersey Institute of Technology (NJIT). On top of that, Dorman carried the Olympic torch through the streets of Los Angeles for the 1984 Olympics.

Albert Dorman carries the Olympic torch through the streets of Los Angeles for the 1984 Olympics

Photo credit: Albert Dorman

After graduating first in his class from NJIT in 1945 with a degree in mechanical engineering, 19-year-old Dorman—who was president of the college student body and editor of the school's yearbook—launched his engineering career by first serving in the U.S. Army Corps of Engineers. Later, he attended the University of Southern California and earned a master's degree in civil engineering while compiling a perfect 4.0 grade point average.

In the late 1940s, Dorman worked for the California State Division of Highways (Caltrans) and City of Los Angeles Department of Building and Safety. During the early 1950s, he was employed by Bowen, Rule and Bowen, a consulting engineering firm.

When he was 28 years old, Dorman formed his own civil engineering company in Hanford, California and, shortly after, also established an architectural-engineering firm in which he was a partner. During this time, he became the civil engineer of record for the one-of-its-kind Disneyland in Anaheim, California (1954–1955).

In 1965, Dorman's engineering firm was acquired by Daniel, Mann, Johnson and Mendenhall (DMJM) in Los Angeles. He rapidly advanced through its ranks until he was president and COO in 1974, then president and CEO in 1977, and chairman and CEO in 1984.

In 1984, Dorman became chairman and CEO of AECOM Technology, which he helped found. He held that position until 1992, when he was named its founding chairman.

Consisting of many of the nation's oldest, largest, and respected employee-owned professional firms, AECOM companies maintain 100 offices and employ approximately 16,000 people worldwide. Dorman has served as chairman of four of the individual firms: DMJM, Consoer Townsend Envirodyne Engineers in Chicago, Frederic R. Harris in New York, and Holmes and Narver in Orange, California.

Under Dorman's leadership, projects AECOM and DMJM have completed globally include mass transit, highways, bridges, tunnels, dams, pipelines, ports and harbors, airports, commercial buildings, wastewater treatment plants, military and industrial installations, and educational, justice, and medical facilities.

Notable projects include Glenwood Canyon I-70 Highway in Colorado; the Air Force One Complex at Andrews Air Force Base and the Baltimore Metro in Maryland; Batman Bridge over the Tamar River in Australia; Port of Sines Breakwater Works in Portugal; Palos Verdes Energy Plant; Delta Airlines Terminal Five at Los Angeles International Airport (LAX); and Los Angeles' Metro Blue, Red, and Green Rail Transit Lines in California.

Born on April 30, 1926, in Philadelphia, Pennsylvania, to William Dorman and Edith Kleiman, Al became a wide-ranging reader as a youngster, enthralled with all fields of human endeavor and especially engineering. Early on, he embraced two of his guiding rules for life: "Always do more than is expected of you and understand the perspectives of those you work for and with."

Neither Al nor his only sibling, younger sister Barbara, followed in their father's footsteps; he owned and operated a rural "general store." Barbara became an attorney and a top executive in the motion picture industry and is currently chair of the UCLA Department of Film, TV and Digital Media. Dorman moved to the highest levels in engineering, architecture, and business, and eventually set his sights on broader vistas.

> "*Civil engineers have a broad responsibility to serve society and make the world a better place. To accomplish this, they must help educate their successors and be ethical exemplars, providing leadership both inside and outside the profession.*"
>
> AL DORMAN

Before starting his engineering education, Dorman spent a year at a liberal arts college. He said, "I wanted to learn about the great ideas in nonengineering fields like philosophy, literature, history, and social sciences. I wanted to mix with people with a wide variety of interests and to learn how best to be able to communicate with them."

On July 29, 1950, Al married Joan Heiten in Buchanan County, Missouri. They would have three children, one daughter and two sons. Laura and Kenneth would become attorneys like their aunt Barbara, and Richard would become a teacher.

Dorman's many activities and leadership roles within the engineering community are too numerous to list and/or describe. A few, however, require mention. He was:

- President of both the Consulting Engineers Association of California (1985–1986) and Los Angeles Section of ASCE (1984–1985)
- Licensed as a professional engineer (PE) in nine states and as an architect in two states, including his adopted home state of 55 years, California
- Author of dozens of papers and lecturer on a wide variety of subjects ranging from "Advanced Transportation Technology," *Science*

Disneyland, Los Angeles, 1955

Walt Disney's brainchild transformed 85 acres of Southern California orange groves into the first modern theme park. It featured a monorail, concealed utilities, and artful landscaping, and became the model for future parks.

Photo credit: Evelyn Weingardt

and Technology (1986 Yearbook) to "The Preservation of Historic Architecture," California Board of Architectural Examiners (1972) and "The Complete Project Manager," ASCE's *Journal of Architectural Engineering* (1995)

• Keynote and convocation speaker at several institutions of higher learning, including Harvey Mudd College, Albert Dorman Honors College, NJIT, California State College at Long Beach, USC, and UCLA

• Trustee, overseer, or councilor at Harvey Mudd College, NJIT, UCLA's School of Architecture and Urban Planning, USC's School of Urban and Regional Planning, and its Board of Councilors for Performing Arts

Along with Dorman's engineering and business achievements came countless professional and public honors, including ASCE's Opal (2000), Parcel-Sverdrup (1987), and Harland Bartholomew (1976) Awards; ACEC's Distinguished Award of Merit (1996); and Los Angeles Area Chamber of Commerce's Herb Nash Award for Economic Development (1988). The University of California at San Francisco presented him with its prestigious Gold Medal in 1996.

In 1998, he was elected to the National Academy of Engineering, and one year later, he received an honorary doctor of science degree from NJIT.

Dorman's extensive pursuits beyond engineering include serving on the board of directors of three major public corporations with revenues ranging into the billions of dollars, and the boards of nine privately held corporations with activities as diverse as product research and development, and agriculture and food processing. He helped establish a savings and loan company and served as its chairman for many years; developed and owned residential subdivisions, apartments, commercial, and industrial properties; and was involved in farming 800 acres in central California.

On the community front, Dorman was, among other things:

• Trustee, The National Foundation for the Advancement in the Arts (1988–1999)
• Director, California Chamber of Commerce (1986–1994)
• Director, LA Area Chamber of Commerce (1983–1988)
• Trustee, The J. David Gladstone Foundation (1988–present)

- Chairman, The Economic and Job Development Committee of the California Chamber of Commerce (1994)
- Director, Californians for Better Transportation (1987–1990)
- President, Kings County Community Concerts Association
- Commissioner, California Department of Conservation, Oil and Gas Division.

Said Dorman, "There is no more noble profession than civil engineering. Whereas medicine frequently focuses on individuals, civil engineering —properly practiced—improves the quality of life for entire communities and populations; safe drinking water, adequate waste disposal, and safe transportation are a few examples."

And because of this, he believes that "civil engineers have a broad responsibility to serve society and make the world a better place." In accomplishing this, "They must help educate their successors and be ethical exemplars, providing leadership both inside and outside the profession. Civil engineering is at the interface between technical engineering and the public; an engineering background and way of thinking can help solve important issues at policy levels as well as at technical levels."

Dorman has walked this talk that he shares with young engineers: "Educate yourself broadly and participate broadly if your career ambitions go beyond a technical specialty."

CHAPTER EIGHT

Educators Extraordinaire

Acquire new knowledge whilst thinking over the old, and you may become a teacher of others.

—Confucius

Teaching, like engineering, is a noble profession. When the two professions are combined and one develops into an engineering professor of high quality, that person attains a significant status to be admired and much respected. When in such a position of influence, that "teacher of others" inspires and motivates students to excel and becomes one of the noblest individuals society can produce.

Without exception, the engineering field is blessed with an abundance of excellent teachers and professors. Not all of them, however, are broad-picture thinkers. Many fail to let engineering students know the importance of their future profession and how it fits into the big picture of society. Engineering is a difficult path to pursue and subjects that must be learned require intense concentration. Learning all the technical subjects necessary and taking enough courses in the humanities to end up with a well-balanced college education cannot be accomplished in four years. The demands that tomorrow's engineers will have placed on them cry out for major changes to be made in today's standard four-year engineering programs, sooner rather than later.

The four featured educators extraordinaire were big-picture sages who not only taught the technical aspects of engineering but also inspired students and framed for them how engineering fit into the whole of society—and how engineers had such a big part to play on the world stage.

The history of formal education for engineers in the United States began with the establishment of the U.S. Military Academy in 1802 under President Thomas Jefferson's watch. Initially set up as a school for engineering officers, the academy's mandate was broadened in 1812 to train officers for all units of the army. Appointed as its first superintendent in 1817, Sylvanus Thayer traveled Europe to study its top schools of engineering, learning the latest in military engineering and the European methods of educating engineers. He then instigated a polytechnic-type curriculum at the U.S. Military Academy. Before long, the academy was graduating engineers with excellent academic credentials.

Wrote Samuel Florman in *The Civilized Engineer*, "West Point graduates, after retiring from the Army, began working as civilian engineers, their numbers increasing from 15 in 1830 to more than 100 in 1838. These ex-military engineers were useful, but the rapidly developing nation clearly needed civilian engineering schools" (Florman 1987).

In response to this need, America's first civilian engineering college was established in 1823. Stephen Van Rensselaer of Albany, New York, an enterprising capitalist, prominent landowner, and community leader, started Rensselaer College in Troy, New York, to teach the application of the physical sciences. Under the direction of Amos Eaton, an engineer, scientist, and lawyer, the school ultimately developed into a professional school of engineering. Its first official degrees in civil engineering were granted in 1835.

In 1849, Rensselaer College was reorganized by Franklin Greene. He had studied and was impressed by European engineering and architectural institutions—in particular the French model—for training engineers and architects. After acceptance of his recommended changes, the school was renamed Rensselaer Polytechnic Institute (RPI). "By 1860, 318 RPI alumni were practicing professionals, along with almost 200 civilian graduates of the U.S. Military Academy," reported Florman. This growing cadre of formally educated engineers along with increasing numbers of on-the-job trained and educated civil engineers (i.e., those produced by the Erie Canal project) were rapidly absorbed by American industry.

Encouraged by this, a number of leading institutions of higher learning followed RPI's and West Point's lead. They instituted engineering programs of their own—Union College in 1845; the Universities of Michigan, Harvard, and Yale in 1847; and then the Massachusetts Institute of Technology (MIT) in 1865.

In the midst of the Civil War, President Abraham Lincoln and the federal government also got involved, approving the 1862 Morrill Act (popularly known as the "land grants" act), which granted federal aid to states establishing colleges with a solid core curriculum in subjects like engineering. "It was the big breakthrough in American engineering education," according to Florman. After the Morrill Act—and the emergence of MIT into the mix—the number of institutions issuing engineering degrees began to swell.

Although many outstanding teachers of civil engineering surfaced early on—and could qualify for a "legends" title—the focus here is on those who emerged in the twentieth century. The first legend is Hardy Cross, whose emergence as a farsighted professor began five decades after the signing of the Morrill legislation. His story, as well as those of Nate Newmark, Mario Salvadori, and Roland Rautenstraus, are being told, not only because of their technical contributions but also because they inspired so many—students, colleagues, and peers—to greatness.

Hardy Cross

Hardy Cross

Photo credit: University of Illinois

"Hardy Cross was the best teacher I ever studied under," said CH2M-Hill cofounder Holly Cornell. Cross, a philosopher as well as an engineer, meshed humanities and engineering ideas in his teachings and inspired numerous engineers to greatness. It was not unusual for Cross to sprinkle his lectures with quotes from the Bible and the classics.

Because he was deaf in one ear, his colleagues and eager students quickly learned which was his better side when they wanted his full attention. It was under his tutelage that the likes of Cornell developed a "great love for engineering." Cross always challenged his pupils to constantly ask, "Why, why, why, and dig until the problem is clearly defined."

As one of the most important names in American civil engineering, Hardy Cross was world famous for his innovations as an educator and a structural theoretician—and for his engineering books and papers, many of which have been translated widely. He was the originator of new methods of structural analysis, many based on using converging approximations. The most significant of these, the fixed-end moment distribution for analysis of continuous structures and indeterminate frames, greatly simplified the way stresses could be calculated.

The moment distribution method, know as the Hardy Cross method, was first published in 1930. Shortly thereafter, Cross developed a similar approach for determining flow in complex pipeline networks.

Essentially, what Cross did was simplify the monumental mathematical task of calculating innumerable equations to solve complex problems in the fields of structural and civil engineering long before the computer age. He revolutionized how the profession addressed complicated problems; whenever engineers in the latter part of the twentieth century talked about methods for designing difficult structures, the name of Hardy Cross was always invoked with awe.

According to Old Dominion University Professor Zia Razzaq (as reported in the *Virginian Pilot* in 1995): "In Hardy Cross's day, if you wanted to design a highway bridge or highrise building, you would end up with several thousand simultaneous mathematical equations. And there were no computers of the kind we have today. He developed a procedure by which, in a very short time, you could actually analyze a very complex structure and calculate all the stresses in it." Without Cross's methods, many engineering projects would have remained dreams and not have become reality when they did (NB-ASCE 2000).

Born on February 10, 1885, in Nansemond County, Virginia, Hardy was the younger of two sons of Thomas H. Cross and Eleanor Wright, both of whom were from prominent southern families. His father, Thomas H., who was attending the University of Virginia when the Civil War broke out, left college to join the Confederate Army. He served in it for the duration of the conflict even though he was wounded during several battles. Shortly before Hardy's older brother, Thomas Peete, was born in 1879, Thomas H. and Eleanor relocated to Norfolk.

Both of the boys were excellent students, and Hardy followed in his older brother's footsteps, graduating from Norfolk Academy and then attending Hampden-Sydney College to become a schoolteacher. He was just 17 when he graduated as valedictorian from Hampden-Sydney with a bachelor of arts degree in 1902. One year later, he received a bachelor of science degree. That same year, his 62-year-old father, who was the local postmaster, died.

Young Cross immediately took a position with his local alma mater, Norfolk Academy, teaching English and mathematics so he could be near his newly widowed mother to comfort her. Three years later, Hardy was found at Massachusetts Institute of Technology (MIT) studying civil engineering. Within two years, in 1908, he had earned an MIT bachelor of science degree.

> "*Engineering does not tell men what they should want or why they want it. Rather it recognizes a need and tries to meet it.*"
>
> HARDY CROSS

Hardy worked briefly as a bridge engineer for the Missouri-Pacific Railway from 1908 to 1909 and then migrated back to Norfolk to teach for one more year. He returned to Boston and, in 1911, received a master of science degree in civil engineering from Harvard University.

Taking an assistant professor's position at Brown University shortly after, he taught civil engineering at Brown for seven years. In 1918, he left the university to become involved in the practice of structural and hydraulic engineering in the Boston and New York areas. During this period, he worked on a number of projects on the East Coast and served as assistant engineer to Charles T. Main.

On September 5, 1921, 36-year-old Hardy married Edythe Hopwood Fenner from Providence, Rhode Island. They would have no children, and she would precede him in death by three years.

The year 1921 also marked the return of Cross to full-time teaching when he took a position as professor of structural engineering at the University of Illinois. By then, he was well known for his insightful writings, including a voluminous report, "River Flow Phenomena and Hydrology of the Yellow River, China," and a popular monograph on the graphic analysis of arch structures, which was being used as a text at Harvard's graduate school of engineering. (They were but two of the many pacesetting engineering publications he would produce over his lifetime.)

In 1937, Cross left the Midwest and returned to the East Coast, accepting his final academic position as chair of the department of civil engineering at Yale, a position he held until his retirement in the early 1950s.

The world-renowned Cross received honorary degrees from Yale (1937), Lehigh (1937), and Hampden-Sydney (1934), and numerous prestigious engineering honors, including ASCE's Norman Medal (1933) and the American Concrete Institute's Wason Medal (1936).

When Cross was named the first recipient of the American Society of Engineering Education's Lamme Medal in 1944, he was cited for "his insistence on the great responsibilities of the individual teacher and his scorn of the superficial in education and for his preeminence in building men who are carrying forward his own high standard for straight, hard thinking in the teaching and practice of engineering."

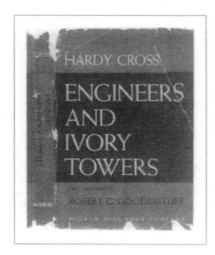

Engineers and Ivory Towers

Tattered dust jacket is testimony to Hardy Cross's masterpiece being an often-read manual. His book, a collection of introspective and thoughprovoking essays about engineering and society, inspired many young engineers to greatness.

Photo credit: William Hall

In his book *Engineers and Ivory Towers,* a collection of his talks, Cross stated that what man wants and what he needs are not always identical. "Engineering does not tell men what they should want or why they want it. Rather it recognizes a need and tries to meet it." His advice to would-be professors was, "One who is to become a teacher of engineering should be trained primarily to be an engineer, and association with the profession outside the ivory towers of learning is absolutely essential" (Cross 1952).

Cross was a lively member of several prominent groups, including ASCE, ACI, American Academy of Arts and Sciences, American Institute of Consulting Engineers, American Railroad Engineering Association, Connecticut Society of Civil Engineers, Royal Society of Arts, and Western Society of Engineers. He was involved in the investigation of several history-altering engineering-construction failures like the Tacoma Narrows Bridge and Charity Hospital in New Orleans. In his later years, he was a much-in-demand speaker and consultant.

In 1958, Cross became the first American to be awarded the Gold Medal of the British Institution of Structural Engineers, where the 73-year-old U.S. engineering superstar delivered the main address at the Institute's fiftieth-anniversary celebration in Manchester. He gave a number of other historic lectures throughout England during the same trip.

Less than a year later, the Benjamin Franklin (scientific) Institute of Philadelphia also selected him to receive its celebrated Gold Medal. It would prove to be a final tribute to a great career and a great engineer, for soon after, on February 12, 1959, the 74-year-old professor quietly passed away in Virginia Beach, Virginia.

Throughout the latter part of his career, Cross often remarked, "People take for granted that an engineer is by definition a technocrat, somebody stumbling across campus with a pen-protector in his front pocket and a satchel full of calculations." The classically educated Cross, however, was far from fitting that mold or any other. He was one of a kind—a far-seeing innovator who always thought outside the box.

Nathan "Nate" Mortimore Newmark

An engineer recognized in part for developing design criteria for tall earthquake-resistant structures and large-scale oil pipelines, Nathan M. Newmark greatly contributed to the U.S. civil engineering profession during his more than half a century of productivity. Along with his earthquake-resistant designs, he was also at the cutting edge in such areas as materials engineering, structural and soil dynamics (both theory and application), numerical methods for analysis, and engineering for national defense.

Nathan "Nate" Newmark
Photo credit: William Hall

Known as Nate, this American engineer developed seismic design criteria for projects such as the Bay Area Rapid Transit System (BART) and the Trans-Alaska Pipeline. His pioneering work in structural dynamics also manifested itself in his seismic analysis of the Latino Americana Tower in Mexico City. The building withstood two large earthquakes, unscathed, in 1957 and 1985. It proved to be a case study in properly designed high-rise buildings successfully surviving major earthquakes.

A year before the 1957 earthquake, Newmark (along with Mexico City consulting engineer Leonardo Zeevaert) had presented a seminal paper on the innovative design of the tower at the World Conference on Earthquake Engineering at the University of California. In their presentation, they justified their design and emphasized the project's unusual characteristics. They said, "The building is nearly twice as tall as any other building in the city, and because of poor foundation soils, a light but rigid structure was designed to rest on a foundation of a floating concrete box set upon piles" (ENR 1957).

So successful was Newmark's seismic analysis that *Engineering News-Record*, in reporting on the 1957 disaster, wrote, "The most encouraging news from earthquake-struck Mexico City is that the city's one true skyscraper, the 43-story Latino-Americana Tower, rode the shock waves undamaged, even to its window glass and partitions" (ENR 1957).

Nate was born on September 22, 1910, in Plainfield, New Jersey, to Abraham S. and Mollie (Nathanson) Newmark. After receiving his early education in North Carolina and New Jersey, young Nate graduated from Rutgers University with high honors and special honors in civil engineering in 1930. He then became a graduate student at the University of Illinois at Urbana, where he received his M.S. degree in 1932, also the year he married Anne May Cohen. Over the years, they raised one son and two daughters, Richard, Linda, and Susan.

He went on to earn his Ph.D. in 1934 and continued as a research assistant, the first of several positions that established his lifelong and rewarding association with this university. His colleague William J. Hall,

Latino American Tower, Mexico City

Designed by Nate Newmark, the Latino American Tower incorporated the latest state-of-the-art seismic engineering techniques, allowing the structural-steel structure to withstand two massive earthquakes, one in 1957, the other in 1985. It came through both occurrences unscathed.

Photo credit: William Hall

Professor Emeritus of Civil Engineering at the University, said Dr. Newmark "carried his own university with him wherever he went, even into professional practice. Engineers, young and old, who came into contact with this man sensed the intellectual and educational challenge. His penetrating insight, his keen engineering judgment, and his genuine interest in people have been a constant source of inspiration to all who have had the privilege of working with him" (Hall 1981).

In 1943, Newmark was appointed research professor of civil engineering and became head of the department of civil engineering in 1956. From 1947 to 1957, he chaired the digital computer laboratory at the university, where he participated in developing one of the country's first large-scale digital computers (ILLIAC II). This achievement marked the beginning of applying computer science to engineering and establishing an entire new department at the university—its digital computer department—which spawned numerous spin-offs and expansion into supercomputing.

Noted Professor Hall, "Professor Newmark possessed unusual ability to attract young people to the field of civil engineering, to inspire them with the confidence to undertake new and varied tasks, to guide but not direct their thinking, and to ensure that as individuals they received appropriate recognition. It is no accident that there grew up around him one of the most active research centers in civil engineering in the country" (Hall 1981).

During Word War II, Newmark served as a consultant to the National Defense Research Committee and Office of Scientific Research and Development. He spent part of his national service in the Pacific war zone. From the mid-1950s onward, he played an important role in developing the design criteria and hardness design of Minuteman and advanced missile systems launch facilities.

Many honors came his way over the years, including the President's Certificate of Merit in 1948.

A founding member of the engineering mechanics division of the American Society of Civil Engineers, Newman won many of the division's awards for individual members and was an ASCE Honorary Member. In 1962, he was elected a Fellow of the American Academy of Arts and Sciences. He was also a founding member of the National Academy of Engineering (NAE) in 1964 and became a member of the National Academy of Sciences (NAS) in 1966.

Two years later, he received the National Medal of Science from President Lyndon B. Johnson, and in 1969, he received the cherished Washington Award (medal), given annually by the Western Society of Engineers with the concurrence of the major U.S. engineering societies. The citation on Newmark's Washington Medal read: "For special contribution to the advancement of engineering knowledge of structures subjected to earthquake or blast and for inspiration to others in improving man's environment."

In the later years of his career, this distinguished engineer received the John Fritz Medal, an all-engineering society award, and the Gold Medal from the Institution of Structural Engineers of Great Britain, becoming only the second American to receive this prestigious award.

"The most encouraging news from earthquake-struck Mexico City is that the city's one true skyscraper, the Latino-Americana Tower, rode the shock waves undamaged"

Engineering News-Record, 1957

When his own university awarded Dr. Newman the honorary degree of doctor of science in 1978, it came with this citation: "Graduate study in structural engineering today bears his indelible imprint as a result of the large group that he attracted to Illinois to work with him. . . . His style, combining rigorous analysis with a sophisticated appeal to experience and intuitive leaps, while inimitable, has provided generations of graduate students with a model of engineering creativity at its best."

Newmark published more than 200 papers, books, and book chapters during his career. He also cowrote *Design of Multi-Story Reinforced Concrete Buildings for Earthquake Motion* with John A. Blume and Leo Corning and *Fundamentals of Earthquake Engineering* with Emilio Rosenblueth.

Indeed, Newmark served his profession and his university in many ways, including the longest appointment to date on the university's research board. In 1973, he became professor of civil engineering and professor in the Center for Advanced Study, taking emeritus status from 1976 until his death on January 25, 1981, in Urbana, Illinois. To seal his legacy, the University of Illinois officially renamed the civil engineering building the Nathan M. Newmark Civil Engineering Laboratory on February 19, 1981.

Mario George Salvadori

Internationally renowned as a brilliant mathematician, inventive engineer, and charismatic professor of structures at Columbia University for more than 50 years, Mario G. Salvadori left a legacy that reached far beyond engineering. "With his boundless engineering knowledge and deep sense of public commitment, he made a unique and wide-ranging contribution to both the University and to society at large," said Kenneth Frampton, a fellow professor at Columbia (Brockway 1997).

Mario Salvadori

Photo credit: Columbia University

Over decades of teaching at universities on two continents, Salvadori inspired countless college-aged students. "His office was considered an ideal training ground for young engineers," reported Frampton. And his inspirational guidance—and fatherly advice—extended even further, into the elementary and junior high school levels.

In the early 1970s, while a partner with the consulting engineering firm of Weidlinger Associates in New York City, Mario began volunteering to educate disadvantaged junior high students in Harlem, mainly about science and engineering. Ultimately these experiences led to the creation of the Salvadori Education Center for the Built Environment, a nonprofit center that supports modern school reform, trains teachers, and educates youngsters. Dozens of inner-city schools in New York continue to use Salvadori's curriculum. One of the main elements of his teaching methods—to get youngsters to fully appreciate science and mathematics—included the use of hands-on study of bridges and other structures. His demonstrations of the applications of engineering principles often included using everyday household products like marshmallows, straws, and toothpicks. His use of folded paper models to create bridges and various kinds of structures always captured the imagination of young and old alike.

"I think Mario would most like to be remembered for the help he gave to inner-city kids who have so much going against them," stated Matthys Levy, a Weidlinger colleague of 40 years. "He treated them like intelligent people and they responded with intelligence. That's his greatest legacy."

Another was his clarity of words. Salvadori wrote 17 books about structural and architectural engineering, many of them crafted so the average person could easily understand complex technical issues. Two of the most popular were *Why Buildings Stand Up* and *Why Buildings Fall Down.* In them, Mario offers his opinion that, more than anything else, "all structural failures may be due to a lack of redundancy" (Salvadori 1980).

Among Salvadori's most well-read technical books were *Numerical Methods in Engineering, Structural Design in Architecture* (coauthored with Levy), and *Structure in Architecture* (coauthored with Robert Heller). About

Structure in Architecture, the famous Italian structural engineer/architect Pier Luigi Nervi (1891–1979) wrote, "Future architects will find it particularly useful to study this book in depth and to meditate upon it, since even if they can entrust the final calculation of a structure to a specialist, they themselves must first be able to invent it and to give it correct proportions. Only then will a structure be born healthy, vital and, possibly, beautiful" (Salvadori 1963).

Both as a teacher and as a practitioner, Salvadori was constantly trying to motivate architects and engineers to work together more closely, pointing out that when they do not, building design suffers. Said James Yao from Texas A&M, "Professor Salvadori criticized the TWA Terminal at Kennedy International Airport because those shell structures were designed by 'brutal force.' They're 'monumental' rather than functional structures, and thus are 'ugly.' How much better they would have been if the architects involved would have better understood structural shell analysis."

Mario wrote in *Why Buildings Stand Up*, "The separation of technology and art is both unnecessary and incorrect; one is not an enemy of the other. Instead it is essential to understand that technology is often a necessary component of art and that art helps technology to serve man better. Nowhere is this more true than in architecture and structure, a marriage in which science and beauty combine to fulfill some of the most basic physical and spiritual needs of humanity" (Salvadori 1980).

Born in Rome, Italy, in 1907, Salvadori received his Ph.D. both in engineering in 1930 and in pure mathematics in 1933 from the University of Rome, where he taught until 1938. In 1940, he fulfilled a persistent dream to emigrate to the United States, where he joined the faculty of Columbia University. While a member of its

CBS Headquarters Building in Manhattan

This building is alleged to be one of the first and most noteworthy reinforced-concrete bearing-wall high-rise structures in the world. Its exterior wall-columns are hollow to contain ducts for the building's HVAC equipment. Clad with imported gray-black Canadian granite, the CBS structure became an instant New York landmark.

Photo credit: Richard Weingardt

School of Engineering and Applied Science, he worked on the Manhattan Project (1942–1945). Later, in 1959, Mario was appointed to Columbia's School of Architecture, Planning and Preservation.

In 1945, Salvadori began a long-lasting collaboration with Paul Weidlinger and his New York-based structural engineering firm. There, he rose to the positions of partner, chairman of the board, and finally, in 1991, honorary chairman. At Weidlinger's, Mario became a world-class expert in thin-shell concrete structures. His work with the Weidlinger group allowed him to extend his classroom design reputation to the real world of construction. Among the significant projects he designed with renowned architects were Walter Gropius's University of Baghdad and Eero Saarinen's CBS building in Manhattan, one of the first reinforced-concrete, bearing-wall high-rises in the world.

> *"It is essential to understand that technology is often a necessary component of art and that art helps technology to serve man better. Nowhere is this more true than in architecture and structure."*
>
> MARIO SALVADORI

In 1997, Mario's wife, Carol, their two sons, Vieri and Michael, and their three grandchildren experienced emotional extremes. In that year, he received the prestigious Founders Award from the National Academy of Engineering (NAE) and, shortly after, he died suddenly at age 90. The NAE honor was one of many prominent recognitions Salvadori received during his productive career. Two others included the 1991 Pupin Medal for outstanding service to the nation in both architecture and engineering and the 1993 Topaz Medallion for excellence in education. He was also an honorary member of the American Institute of Architects.

Salvadori remained active professionally until the end of his long life. His book about earthquakes and volcanoes, *Why the Earth Quakes*, was published when he was 88 years old. At the time of his death, he was still active at Columbia as the James Renwick Professor Emeritus of Civil Engineering and a professor emeritus in the School of Architecture, Planning and Preservation.

Salvadori will always be remembered as an inspirer of young people and a designer of leading-edge structures. He was also one of America's most enthusiastic champions linking the fields of structural engineering and architecture for the betterment of both professions. In 1999, he was named one of *Engineering News-Record*'s top 20 structural engineers of the last 125 years.

Roland Curt "Raut" Rautenstraus

Called American's version of Britain's Alistair Cooke, he was the man with the golden voice. In his long, illustrious, and inspirational career, Roland C. Rautenstraus—an eminent professional engineer—was also teacher, role model, and leader to several thousand young men and women who crossed his path. He stirred them to reach deep inside and become greater than they ever thought possible.

Stan Adelstein, a successful engineer and contractor, who is currently a South Dakota state legislator, acknowledged that Professor Rautenstraus (Raut) was his inspiration. "He could make a person look at things differently. He had the magic ability to paint a picture with words—almost like music that you don't forget. Raut could reach into the subconscious and plant ideas in such a positive way, you could always hear—and visualize—them."

Roland Rautenstraus

Photo credit: Curt Rautenstraus

Another one of Roland's students, Kristy Schloss, stated, "Raut encouraged us to be independent thinkers, to not only define the problems but also to determine how best to solve them. He excited us and inspired us. He got us hooked into wanting to be the best we could be." Kristy currently heads a 100-year-old international environmental equipment design and manufacturing company. Like thousands of other engineering leaders who graduated from the University of Colorado, she credits Raut for her success.

Rautenstraus was the only professional engineer (PE) ever elected president of the University of Colorado, considered by many as the most prestigious university in the Rocky Mountain region. He reached that lofty position after several years of teaching, then serving as chair of civil and architectural engineering and as vice president of the university. He ascended to the presidency in 1974 during the troubled times after the end of the Vietnam War. Raut not only brought calm to the situation but also expanded the university from a single-campus operation to a multicampus institution of higher learning.

About Rautenstraus's reign from 1974 to 1980, Raut's boyhood (and lifelong) friend former U.S. Senator Bob Dole of Kansas said, "His years of overseeing the institution as its president helped the University acquire international stature. Because of his ability to see the big picture, his contributions to education and the profession of engineering were significant and visionary. A credit to Colorado and the University, Raut enhanced the national reputation of both" (Weingardt 1999).

Senator Dole also stressed Rautenstraus's impact internationally saying, "In his later years [after 1980], he was an international engineering consultant and advisor for businesses in developing countries. A recognized leader in this field, he gave lectures worldwide on how to feed and industrialize third world countries through technology transfer." His legacy in this area, especially in Central America, will extend well into the twenty-first century.

Journalist Edward Murray, who interviewed Rautenstraus in 1977 halfway through his presidency, wrote, "One of the joys of interviewing Rautenstraus is that he sounds as if he just finished a day's research on whatever subject comes up. He doesn't lecture or spout facts; he targets in with the concise information most pertinent to the question. At age 53, Roland looks like a movie star, thinks precisely like the civil engineer he is, and talks like an ethical philosopher. It's no wonder he is president of the state's elite university with its 30,000 students, its 1,973 faculty members and 6,309 supporting staff, its four campuses, and its $250 million annual budget. He exudes the competence and confidence to handle the job" (Weingardt 1999).

> *"Don't live in the gray monotone world of attempting to make no errors. If you do, you'll end your lives wondering if you ever lived at all—and that would be the greatest of all mistakes."*
>
> ROLAND RAUTENSTRAUS

Roland was born on February 27, 1924, in Gothenburg, Nebraska, the third and youngest son of Reverend Christian and Mrs. Emma Rautenstraus, who were 1908 Lutheran immigrants from Wurtenburg, Germany. Roland's favorite sibling Ruth, 10 years his elder and at one time his grade school teacher, both coddled and challenged her youngest brother to develop his talents to the highest. Raut was the youngest by 10 years in a caring family, which offered a lot of advantages. According to Rautenstraus, "There were always a lot of adults around who paid attention to me and taught me things." On a continual basis he conversed with people older than himself, fraternized with them, played adult card games, learned to read young, and, as a result, grew early.

After Gothenburg, Christian was assigned to Creston, a small rural parish in eastern Nebraska, where the family experienced the terrible effects of the Great Depression.

When he was in his 40s, Rautenstraus appeared as the main speaker on a June 1966 Boulder, Colorado, TV special, "The Depression Years." He recapped the Depression years saying:

> My father went three years without getting a salary, so we were devoid of any money in the family. But we weren't alone. Everyone back then pulled together. Every family member, even young kids, took on any odd job they could get and neighbors helped neighbors, and it brought people closer together. Those things reflected the spirit and the ingenuity of Americans during the Depression, but the times were far from great. Some monstrous human catastrophes occurred. Farmers, right and left, lost their farms. Some people who had to declare bankruptcy shot themselves. It was a style of life so alien from what we know today that it's hard to conceive it even happened in the same country where we now live (Weingardt 1999).

In 1934, Reverend Rautenstraus became the pastor of St. John's Lutheran Church in Russell, Kansas, which is where Roland became friends with Bob Dole. Both of these handsome young men excelled in sports and academics. As the Kansas state debate champion in high school for two years in a row, Raut began his road to becoming a world-class speaker.

As described by David Grimm, his presidential assistant at the University of Colorado from 1974 to 1980, Rautenstraus eventually became "an orator of the highest order, right up there with the best in the world. He had a fantastic and unique style and made a lasting impression, not unlike John F. Kennedy or Winston Churchill" (Weingardt 1999).

"Picture frame" bridge on Genesee Mountain

The bridge crosses I-70 and frames the rugged Rocky Mountains. In addition to being a member of Eisenhower's interstate advisory team, Rautenstraus had a major impact in modernizing Colorado's entire highway system.

Photo credit: David R. Weingardt

Shortly after the Japanese bombed Pearl Harbor in 1941, Roland enrolled at the University of Colorado in the U.S. Navy's officers training program as an engineer. His father had wanted him to be a teacher or a minister, but the Navy needed engineers, so for one of the few times in his life, he went against his father's wishes. At the university, he starred on its football team while taking a full course load to graduate in three years.

In 1946, after serving with the U.S. Navy Seabees in Okinawa, Roland married his college sweetheart, Willy Atler. Willy became a teacher, and Roland—hired by his former engineering dean—became an engineering professor. By the time their only child, Curt, was born in 1947, Rautenstraus was heavily involved with national engineering committees for the American Society of Surveying and Mapping and several other professional groups. He also received crucial public appointments, including one on President Eisenhower's national advisory committee for the interstate highway system. He also served on numerous local civic boards, like the Governor's Energy Task Force, Colorado Council on Arts and Humanities, Denver Center for the Performing Arts, and Boulder Water Board and Planning Commission.

Among Rautenstraus's favorite awards and honors for lifetime achievements as an engineer and/or leader were the University of Colorado's two highest—the Robert Stearns Award (1965) and the Norlin Medal (1974). He is the only person in history to be the recipient of both. Also high on his list of favorites were an honorary doctor of law degree from the University of New Mexico (1976) and the General Palmer Award from the American Council of Engineering Companies of Colorado (1996).

For 50 years—from 1947 to 1997—Rautenstraus, as teacher, leader, and engineer, profoundly touched the lives of countless University of Colorado engineering students, colleagues, and many others, both within and beyond the state. His passing on Christmas Day, 1997, brought many eulogies. None was more fitting than one by Vince Kontny, a former student and president of Fluor Corporation: "Raut made students—and practitioners—proud of the profession. He showed how engineering fit into the big picture of things. What he taught me I used in the real world."

As a much-sought-after public speaker his whole life, the golden-voiced Rautenstraus frequently delivered major keynote addresses, often at college commencement ceremonies around the world. He always challenged his listeners, especially young future leaders, to strive to be the best they could—and *not* to hold back. One of his favorite lines was, "Don't live in the gray monotone world of attempting to make no errors. If you do, you'll end your lives wondering if you ever lived at all—and that would be the greatest of all mistakes" (Weingardt 1999).

CHAPTER NINE

Tomorrow's Legends

Leaders are made not born. Today, we desperately need women and men who can take charge, and we must raise the search for new leadership to a national priority.
—Warren Bennis

Studying the lives and derring-do of history-shaping American engineering icons such as those collected in *Engineering Legends* leads to three clear observations.

The first is captured by Henry Petroski, who said, "What makes history interesting and relevant is it not only teaches us about the way things used to be done; it also gives us perspective on how things are done today—and how they most likely will be done in the future" (Weingardt 2004). And as Thomas Carlyle reminds us, "History is but the biography of great men."

The second is that the noble profession of civil engineering encompasses a wide range of specialties and activities, and that those who reach greatness in it can come from anywhere, socially and geographically. These greats excel in both the private and public sectors and as members of small or large companies and/or concerns. Engineering legends are testimony to famed football coach Vince Lombardi's statement: "The quality of a person's life is in direct proportion to their commitment to excellence, regardless of their chosen field of endeavor."

The third observation: Ever since the founding of the United States, leading civil engineers have been key in contributing to its progress. They have been involved in the creation and building of the nation's modern facilities, advanced engineering systems, and technological marvels. What civil engineers have done and continue to do adds value and makes every citizen's daily life more satisfying and productive. The products of an engineer's work are uplifting to the human spirit.

Because of that, top U.S. civil engineers will continue to be in huge demand, even more in the future than in the times when this book's superstars lived. To rise to the demands of today's current world order and global economy, however, nascent engineering legends will be called upon to not only match but also exceed the exploits, innovations, and accomplishments of their predecessors.

Just being a technically competent engineer only able to do engineering will not be enough.

Tomorrow's civil engineers will have to be both leading-edge technical experts and world-class leaders. The examples set by the visionary and pioneering civil engineering giants of days gone by provide superb guidance. Along with having defined the profession as it is now known, they have raised the engineering community's collective consciousness of what can be achieved. They have laid out a road map for today's engineers to follow and raise the bar of excellence even higher.

Dependence on Engineering

It is increasingly clear that science, engineering, and technology will advance more over the next few decades than in all recorded history, and that the skills of every nation's engineers—and, indeed, the scientific and technological literacy of its populace in general—will be put to the test, especially in the United States as the world's only remaining superpower.

As the world moves deeper into the twenty-first century and becomes increasingly dependent on sophisticated technology, modern society will more than ever be required to put trust in its machines and technical problem-solving experts—and rely on them *not* as followers but as leaders in positions of consequence. To be effective at helping set sound public agendas and policies, civil engineers, in particular, will have to be well-rounded professionals, rather than narrowly focused technicians.

To suitably resolve the burgeoning environmental, infrastructure, and sustainable development problems earth will experience in the coming years, twenty-first-century civil engineers will not have the luxury of solely filling purely technical roles. To maximize their impact on civilization's progress, civil engineers of the future must boldly step forward as societal leaders and effective decision makers in public affairs and even in politics.

The importance of civil engineers becoming involved beyond engineering is underscored by Bernstein and Lemer in *Solving the Innovation Puzzle: Challenges Facing the U.S. Design and Construction Industry*: "The U.S. today possesses a physical infrastructure of extraordinary scale and scope. This civil infrastructure supports virtually all elements of our society, and the people and businesses that have produced it comprise a major segment of our economy. History indicates that the growth, flourishing and decline of any civilization are closely mirrored by the life cycle and performance of its civil infrastructure" (Bernstein 1996).

As has been true in past history, engineering and technology will lead the most progressive changes. To control their own destinies, more and more civil engineers must become involved in setting strategy and direction in the broader community. Said Bernstein and Lemer, "In the areas of technological development change will come fastest; it will come more slowly in political and social arenas. Paradoxically, it is in this area of social concerns about engineering and construction in the built environment that industry's professionals and leaders can most increase their influence."

Improving how favorably the public perceives civil engineers and their profession will similarly affect their performance, both as technical experts and as leaders. More than that, it will greatly influence whether civil engineering is thought of as a commodity and its members as professionals on a par with doctors, lawyers, and scientists or as technicians to be hired solely by low cost.

> *"The products of an engineer's work are uplifting to the human spirit."*
> RICHARD WEINGARDT

Although the overall need worldwide for people with strong engineering backgrounds will continue to multiply—as will the potential for U.S. civil engineers to become a powerful force in society—there is great concern about American youngsters turning away from the profession in droves. Too few of them have been made aware of engineering's relevance in their lives going forward and the legendary engineers that added that relevancy.

Global Demand

For a number of reasons, the global demand for civil engineering services will increase dramatically in the coming decades. Among them are an increased dependence and reliance on engineering to successfully:

- Incorporate space-age advancements in engineering and technology to improve everyone's daily lives.
- Come up with sound, practical solutions for sustainable development and a healthier planet.
- Explore space, the oceans, and the earth's core.
- Deal with the world's exploding population.

By the end of the twenty-first century, the United Nations predicts that the world's population will have reached its maximum level—10 billion people. America's population is expected to double over the same time span. Interestingly, this is in direct contrast to forecasts for other industrial countries, including Europe and Russia. Their populations are actually predicted to decline in this century (Snyder 2001).

Because of the global population trends, infrastructure development, replacement, and/or renewal will likely constitute the majority of future civil engineering work on all corners of the world. This will take place more in developing countries than in developed ones, except for in America.

Doubling America

The doubling of America's population means this country needs to double the size of its national infrastructure as well as upgrade and repair its existing public facilities. And a massive building expansion by the private sector will have to accompany or lead extensive public sector development.

Large numbers of civil engineers, in general, will be needed to build the endless number of projects required to double the "size" of America. As much as any other factor will be the tremendous need for top-notch civil engineering pacesetters and innovators to make these ventures as efficient as possible. Plus, their presence will be needed to lead changes in society because the construction of an "entire additional America" will have to occur within increasingly stringent environmental and land use constraints. This, by itself, will pose a substantial civil engineering challenge over and above the actual technical work.

The call to deal with nonengineering issues—and public concerns about the impact that massive increases in the built-environment will have—will put civil engineering on center stage. Consequently, the demand for creative engineers will magnify exponentially. Three big questions, however, must be answered and addressed by today's U.S. civil engineering community to be able to optimize these future opportunities for engineering services. They are:

1. How much of this much-needed civil engineering work for U.S. projects will be done by U.S. engineers and how much by lower-salaried, non-U.S. engineers?
2. On how much of this work will civil engineers be hired by low-bid versus Quality Based Services (QBS—the federal U.S. Brooks Bill) procedures?
3. How much of this work will be automated on computers without need for human attendance?

The outstanding examples set by past engineering visionaries—along with their "pearls of wisdom" and sage advice—will help in thinking through resolution of these issues and other dilemmas on the horizon. Just as past legends had their defining moments, so will this generation's trailblazers.

Public Surveys

In poll after poll, public surveys consistently confirm that engineers are held in high esteem, respected for their ethics and honesty. The public basically perceives that civil engineers are smart, hardworking, and trustworthy. It is just that the majority of people do not know exactly what engineers do! Or why they are so needed for progress and the creation of wealth!

Plus few, if any, of the profession's top engineers have the same public name recognition as superstars in other professions, for example, Frank Lloyd Wright and Frank Gehry in architecture, Johnnie Cochran and Gerry Spence in law, Albert Einstein and Marie Curie in science, and Jonas Salk and Louis Pasteur in medicine.

The message about engineering being an exciting and history-making profession has been long overlooked. That oversight drives home the point that if the profession will ever escape from invisibility, civil engineers as a group must step forward and be heard. It is such a loss when names of engineers so key in changing the course of history and civilization disappear. If that continues, America's young will never get to experience the excitement of their exploits.

> *"The quality of a person's life is in direct proportion to their commitment to excellence, regardless of their chosen field of endeavor."*
>
> VINCE LOMBARDI

It would behoove civil engineers of tomorrow to become masters at dealing with the media, convincing them of the awe-inspiring significance of civil engineering designs. The time has also come for a centralized national civil engineering museum showcasing engineering achievements and projects, and incorporating a Hall of Fame for Civil Engineers. The legends featured and cited here could be the first inductees.

Attracting America's Youth

Today, when bright young people who have the talent to become anything they want consider their life's work, they look for role models. They seek heroes and heroines within the fields they consider entering. They search for individuals they can look up to and connect with.

Who are the contemporary heroes and role models American youngsters are exposed to? As it stands, they are inundated by mainstream media, movies, and TV shows displaying the antics of celebrities, mostly entertainers, many of whom have dubious credentials and moral values.

Surely typical media-propagated celebrities are not the only—or best—role models for youth in the long run. How much better it would be if America's young were exposed to a diversity of role models and superstars, including modern-day Thomas Edisons and Henry Fords, and civil engineering legends like those in this book—heroes and heroines who have deep convictions and integrity, who display solid work ethics.

Countless civil engineers with the right stuff to be universal role models live in the United States. They are role models not only for young engineers but also for all Americans, no matter what their calling. It is time to name as many of them as possible and place them in the public spotlight along with past legends.

It is unlikely that the subjects of media reports will change any time soon. However, that need not discourage or prevent engineers from doing their part to introduce today's youth to a broader range of hero figures than film, rock, and sports stars. That is why the engineering community must take the bull by the horns and instigate a comprehensive, unified public awareness campaign that would make it a national priority to identify and celebrate its stars by promoting stories of civil engineering feats, especially to the youth of the country.

> "*What makes history interesting and relevant is it not only teaches us about the way things used to be done; it also gives us perspective on how things are done today—and how they most likely will be done in the future.*"
>
> HENRY PETROSKI

One way to accomplish this would be by maximizing the potential of the Internet to celebrate the profession. Another would be to produce more books that showcase engineers, written not in the "language" of engineering but in a language understandable to non-engineers. These books could be colorful coffee-table books on outstanding engineering projects that have relevance and engineers as people, similar to those the architectural community publishes about their projects and stars.

A Century of Innovation: Twenty Engineering Achievements That Transformed Our Lives, produced by the National Academy of Engineering in 2003, although not exclusively about civil engineering, is one example of the types of coffee-table books needed.

Wide distribution of novels with civil engineers in leading roles would also be important in a nationwide awareness campaign—novels like Samuel Florman's *The Aftermath* and Robert Byrne's *Skyscraper.* They, *Engineering Legends*, and books clearly featuring the heroic deeds of engineers, like David McCullough's *The Great Bridge* and *Path between the Seas*, should be in every high school library in the country.

Bemoaning the fact that there are few movies made about engineers or that there is no "L.A. Engineering" TV series portraying the wonderfulness of engineering, accomplishes little. Taking action and becoming involved in efforts to see that such happens is the only way to correct this.

There are many engineering excellence awards programs in existence, but none, unfortunately, captures the attention of the public or media, certainly not like Hollywood's Academy Awards or awards for sports heroes. And, even though it may be impossible to ever garner the level of exposure those programs get, certainly a dream profession like engineering can come up with some awards that better capture the media's attention than has been done in the past. An award named the "Leonardo da Vinci Medal," recognizing multifaceted civil engineering legends in the industry—Renaissance men and women—might be an answer.

Tomorrow's Legends

Who will be tomorrow's civil engineering legends? Those who are in the heat of the battle addressing such issues as productivity, the environment, and sustainable development—and defining where the U.S. civil engineering profession will head in the future.

There has never been a shortage of civil engineering masters to spotlight. Not yesterday, not today, not tomorrow, not by a long shot! For instance, in a nationwide survey conducted through *Structural Engineer* magazine in June 2003, engineering leaders were asked to name America's most influential and/or powerful living (at the time) structural engineers (Weingardt 2003).

Three of the categories in the survey were education, design, and innovation. With countless names submitted in each, the following engineers topped the list of vote-getters.

- *Education.* Louis Geschwinder, Frieder Seible, Lynn Beedle, David Billington, Jack McCormac, William Hall
- *Design.* Leslie Robertson, Charles Thornton, William LeMessurier, Matthys Levy, Man-Chung Tang, Lawrence Griffis, Nabih Youssef
- *Innovation.* John Fisher, Omer Blodgett, Shankar Nair, George Housner, Horst Berger

Many of these, if not already so recognized, will officially be elevated to structural engineering legendhood in due course. Many other engineers practicing different disciplines of civil engineering will also be noted. Their identities could quickly be revealed through simple surveys; one among the reader's peers and through industry publications similar to the one conducted in *Structural Engineer*.

There are several contemporary American civil engineering luminaries who are legends today or soon will be: Hank Hatch, Henry Michel, Eugene Figg, Ben Gerwick, Fred Heger, Blair Birdsall, Mary Cleave, Charles Pankow, Samuel Florman, Ralph Peck, Daniel Okun, Joseph Ling, Marilyn Reece, Ross McKinney, Malcolm Pirnie, Margaret Petersen, Herb Rothman, Julia Weertman, James Geraghty, and Kenneth Wright. What other noteworthy civil engineers meet the requirements for legendhood? There are many! Compile your list today, and then make it known to colleagues, media representatives, and the public in general.

The Beginning

Engineering Legends represents a new beginning. Although it recaps the past by identifying several legends from yesteryear, its message looks forward to the future. It gives food for thought on how civil engineers will, can, and should fit into the world of tomorrow by knowing how past legends have influenced history. This book calls all engineers to action, challenging them to reach for greatness and be the best they can be, both as leading-edge engineering pioneers and as citizens of the world. It opens eyes to the possibilities available in this profession so all engineers can maximize their talents and make the world a better place than they found it.

For much too long, civil engineers as a whole have been silent and invisible—overly shy and humble. A better rallying cry than "Do good work and hope someone appreciates it" would be "Do good work and tell the world about it." For that to happen, increasing numbers of civil engineers must get out of their shells and communicate with large numbers of people outside the narrow confines of the engineering industry. They must leave their own legacies of inspiration and achievement to show the way for future generations.

The world is run by those who show up, and the need for more and more civil engineers to show up in "big-picture" policy and decision-making positions is more critical than ever. If civil engineers do not take on meaningful leadership roles to shape the public agenda in the areas where only engineers possess the necessary expertise, they will have abdicated their rightful role and responsibility to properly address technological and engineering outcomes. They will leave them to be dealt with by those less than qualified.

Shaping the Future

The dawning of the twenty-first century finds mankind at the crest of one of the most creative and exciting periods in history. The potential for innovative civil engineers to significantly shape tomorrow's built environment and set the standards for future generations is awesome. By building on the work of legends before us, this generation of civil engineers can leave a legacy that is even greater.

Being a technically skilled engineer who helps create wealth by enlarging the economic pie and protects everyone's standard of life will, in and of itself, be rewarding. And for those civil engineers who become "movers and shakers" in industry and/or society the future will be even be more exhilarating.

Acronyms

AAE—American Association of Engineers
AAES—American Association of Engineering Societies
ACI—American Concrete Institute
ACEC—American Council of Engineering Companies
AIA—American Institute of Architects
AISC—American Institute of Steel Construction
ASCE—American Society of Civil Engineers
CCOM—ASCE's Communication Committee
CEC—Colorado Engineering Council
CEO—Chief Executive Officer
COO—Chief Operations Officer
DIA—Denver International Airport
DRB—dispute review board
EERI—Earthquake Engineering Research Institute
ENR—*Engineering News-Record*
FERA—Federal Emergency Relief Administration
FIDIC—International Federation of Consulting Engineers
FT, ft—foot, feet (ft = 0.3048 meters)
IN, in.—inch, inches (in. = 25.4 millimeters)
KM, km—kilometer, kilometers (km = 0.63 miles)
LAX—Los Angeles International Airport
M, m—meter, meters (m = 3.28 feet)
MIT—Massachusetts Institute of Technology
MGD—millions of gallons per day
NAE—National Academy of Engineers
NAS—National Academy of Science
NASA—National Aeronautics and Space Administration
NJIT—New Jersey Institute of Technology
NSF—National Science Foundation
NSPE—National Society of Professional Engineers
OSC—Oregon State College
PCA—Portland Cement Association
PCI – Prestressed Concrete Institute

PE, P.E.—professional engineer
PSI, psi—pounds per square inch
QBS—Quality Based Selection (Federal Brooks Bill)
SF, sf—square foot, square feet
SOM—Skidmore, Owings & Merrill
RIT—Rochester Institute of Technology
RPI—Rensselaer Polytechnic Institute
RWC—Richard Weingardt Consultants, Inc.
TMI—Texas Military Institute
TWA—Trans-World Airlines
U.K.—United Kingdom
UN—United Nations
USBR—U.S. Bureau of Reclamation
WSE—Western Society of Engineers

References

Ali, Mir. 2001. *Art of the Skyscraper: The Genius of Fazlur Khan.* Champaign, Ill.: Rizzoli Publishing.

Ambrose, Stephen. 2000. *Nothing Like It in the World.* New York: Simon & Schuster.

AME. 1960. "David B. Steinman," *American Engineer Magazine* (Sept.), V. 30, N. 9, NSPE Publisher.

Bartels, Nancy. 1997. "The First Lady of Gearing," *The Journal of Gear Manufacturing* (Sep./Nov.).

Baume, Michael. 1967. *The Sydney Opera House Affair.* Camden, N.J.: Thomas Nelson and Sons.

BBC. 2001. "Leonardo Bridge Opens 500 Years Late," *BBC News* (Oct. 31), pg. 1.

Bennett, Helen. 1928. "Kate Gleason's Adventures in a Man's Job," *The American Magazine* (Oct. 28), pp. 42–43, 158–175.

Bennis, Warren. 1989. *On Becoming a Leader.* New York: Addison-Wesley Publishing Company, Inc.

Bennis, Warren. 1990. *Why Leaders Can't Lead: The Unconscious Conspiracy Continues.* San Francisco, Calif.: Jossey-Bass Publishers.

Bernstein, Harvey, and Andrew Lemer. 1996. *Solving the Innovation Puzzle: Challenges Facing the U.S. Design and Construction Industry.* Reston, Va.: ASCE Press.

Bernstein, Harvey. 2000. "The Future of Engineering," *Eye to the Future,* pg. 45. Washington, D.C.: ACEC Publications.

Bey, Lee. 1998. "Chicago's Towering Intellect: Fazlur Khan," *Chicago Sun-Times* (June 15).

Billington, David. 2003. *The Art of Structural Design: A Swiss Legacy.* Princeton, N.J.: Princeton University Art Museum.

Billington, David. 1996. *The Innovators: The Engineering Pioneers Who Made America Modern.* New York: John Wiley & Sons, Inc.

Billington, David. 1983. *The Tower and the Bridge.* New York: Basic Books, Inc.

BPP. 1998. "RIT Receives $10M from Gleason Foundation," *The Brighton-Pittsford Post* (Jul. 8).

Brockway, Kim. 1997. "Mario Salvadori, Architect, Engineer," *Columbia University Record* (Sept. 12), V. 23, N. 2.

Burke, James. 1978. *Connections*. Boston, Mass.: Little, Brown and Company.

Byrne, Robert. 1984. *Skyscraper*. New York: Atheneum.

CE. 1937. "John Alexander Low Waddell (His Life)," *Civil Engineering* (Jan.), pg. 77.

CE. 1997. "What's Bill Hall Doing with Hardy Cross's Wallet?" *Civil Engineering* (Fall), University of Illinois at Urbana-Champaign.

Chanute, Octave. 1976. *Progress in Flying Machines* (Reprint). Long Beach, Calif.: Lorenz and Herzog Publishers.

Chappell, Eve. 1920. "Kate Gleason's Careers," *The Women Citizen* (Jan.), pp. 19–20.

Chen, Fu Hua. 1996. *Between East and West*. Boulder, Colo.: University Press of Colorado.

CM. 1966. "Soldier Beams and Concrete Float Building in Weak Ground," *Construction Methods* (Mar.).

Colvin, Fred. 1947. *60 Years with Men and Machines*. New York: Whittlesley House.

Condon, George. 1974. *Stars in the Water: The Story of the Erie Canal*. Garden City, N.Y.: Doubleday & Company, Inc.

Constable, George, and Bob Somerville. 2003. *A Century of Innovation: Twenty Engineering Innovations that Transformed Our Lives*. Washington, D.C.: Joseph Henry Press.

Cross, Hardy. 1952. *Engineers and Ivory Towers*. N. Stratford, N.H.: Ayer Company Publishers, Inc.

Duis, Perry. 1976. *Chicago: Creating New Traditions*. Chicago, Ill.: Chicago Historical Society.

Dumas, Alan. 1996. "700 Miles of Bad Memories," *Rocky Mountain News* (Sept. 22), Section D, pp. 15–17.

Dupre, Judith. 1996. *Skyscrapers*. New York: Black Dog and Leventhal Publishers.

Eaton, Leonard. 2001. "Hardy Cross and the 'Moment Distribution Method'," *Nexus Network Journal* (Summer), V.3, N.3.

ENR. 1957. "No Damage to Tallest Building." *Engineering News-Record* (Aug.), pp. 9–12.

ENR. 1999. "Top People of the Past 125 Years." *Engineering News-Record* (Aug.), pp. 27–54.

Faber, Colin. 1963. *Candela: The Shell Builder*. New York: Reinhold Publishing Corporation.

Fellerman, Hazel. 1936. *The Best Loved Poems of the American People*. New York: Garden City Books.

Fincher, Jack. 1905. "George Ferris, Jr. and the Great Wheel of Fortune," *Smithsonian* (July), pp. 109–118.

Fleischman, Doris (ed.). 1929. *An Outline of Careers for Women: A Practical Guide to Achievement*. Garden City, N.Y.: Doubleday, Doran and Company, Inc.

Florman, Samuel. 2001. *The Aftermath*. New York: St. Martin's Press.

Florman, Samuel. 1987. *The Civilized Engineer*. New York: St. Martin's Press.

Foss, Sam. 1895. *Whiffs from Wild Meadows*. Boston, Mass.: Lothrop, Lee and Shepard.

Fox, Robin Lane. 1980. *The Search for Alexander*. Boston, Mass.: Little, Brown and Company.

Fredrich, Augustine (ed.). 1989. *Sons of Martha: Civil Engineering Readings in Modern Literature*. New York: ASCE Press.

Freeman, Michael, et al. 2003. "Past, Present, and Future of the Wood Preservation Industry," *Forest Products Journal* (Oct.).

Fuller, Craig. 1992. "George Dewey Clyde," *Beehive History*, pp. 25–27.

Fuller, R. Buckminster. 1963. *Ideas and Integrities: A Spontaneous Autobiographical Disclosure*. Englewood Cliffs, N.J.: Prentice-Hall.

Fuller, R. Buckminster. 1969. *Operating Manual for Spaceship Earth*. Carbondale, Ill.: Southern Illinois University Press.

Funderburg, Anne. 1993. "America's Eiffel Tower," *Invention and Technology* (Fall), pp. 9–16.

Galloway, John. 1941. "Theodore Dehone Judah—Railroad Pioneer (Parts I and II)," *Civil Engineering* (Oct. and Nov.), V. 11, N. 10 and 11.

Gerhardt, Manfred. 1984. "Some Recollections of the Engineering Experiences of Willard E. Simpson," (White Paper Prepared by Gerhardt as an Officer of the Simpson Group), San Antonio, Tex.

Goldberger, Paul. 1981. *The Skyscraper*. New York: Alfred A. Knopf, Inc.

Goodman, Richard. 1999. *Karl Terzaghi: The Engineer As Artist*. Reston, Va.: ASCE Press.

Gorn, Michael. 1992. *The Universal Man. Theodore von Karman's Life in Aeronautics*. Washington, D.C.: Smithsonian Institution Press.

Gottlieb, Agnes Hooper, et al. 1998. *1,000 Years, 1,000 People: Ranking the Men and Women Who Shaped the Millennium*. New York: Kodansha America, Inc.

Gray, Carl. 1927. *The Eighth Wonder*. Boston, Mass.: B. F. Sturtevant Company.

Griggs, Francis. 2003. "1852–2002: 150 Years of Civil Engineering in the USA," *Perspectives in Civil Engineering*. Reston, Va.: ASCE Press.

Griggs, Francis. 2004. "Timothy Palmer: The Nestor of American Bridge Builders," *STRUCTURE Magazine* (July), pp. 34–35.

Gtomb, Jozef. 1981. (Translated into English by Peter Obst, 2002.) *A Man Who Spanned Two Eras: The Story of Bridge Builder Ralph Modjeski*. Philadelphia, Penn.: Kosciuszko Foundation (Philadelphia Chapter).

Gullan, Harold. 2004. *First Fathers: The Men Who Inspired Our Presidents*. Hoboken, N.J.: John Wiley & Sons, Inc.

Hall, William. 1981. "Nathan Mortimore Newmark (A Memorial)," *Bulletin of the Seismological Society of America* (Aug.), V. 71, N. 4.

Hansen, James. 1995. *Enchanted Rendezvous: John C. Houbolt and the Genesis of the Lunar-Orbit Rendezvous Concept* (Monographs in Aerospace History – No. 4). Washington, D.C.: NASA History Office.

Hauck, George, and Louis Potts. 1996. "J. A. L. Waddell and the Diffusion of Civil Engineering Techniques," *Civil Engineering History: Proceedings of the First National Symposium on Civil Engineering History*. ASCE.

H&H. 1987. "One Hundred Years of Bridge Engineering: 1887–1987," *Hardesty and Hanover Brochure*, New York.

Hoddeson, Lillian (editor). *No Boundaries*. Urbana, Ill.: University of Illinois Press.

Huxtable, Ada Louise. 1982. *The Tall Building Artistically Reconsidered: The Search for a Skyscraper Style*. New York: Pantheon Books.

Iovine, Julie. 2004. "A Tiny Museum Achieves Its Towering Ambition," *The New York Times* (Feb. 29), Section 2, pg. 28.

Jablow, Valerie. 1999. "Othmar Ammann's Glory," *Smithsonian Magazine* (Oct.).

Jackson, Donald. 1988. *Great American Bridges and Dams*. Washington, D.C.: Great American Places Series, The Preservation Press.

Johnson, Paul. 1999. *A History of the American People*. New York: Harper Perennial.

Kass-Simon, Gabriele and Patricia Farnes (eds.). 1990. *Women of Science*. Bloomington, Ind.: Indiana University Press.

Kennedy, John F. 1955. *Profiles in Courage*. New York: Harper and Brothers.

Khan, Yasmin. 2004. *Engineering Architecture: The Vision of Fazlur R. Khan*. New York: W. W. Norton & Company.

King, John. 2003. "Tung-Yen Lin, 1912–2003: Renowned Engineer Helped Shape Bay Area and the World, " *San Francisco Chronicle* (Nov. 18).

Kirby, Richard, et al. 1956. *Engineering in History*. New York: McGraw-Hill Book Company, Inc.

Larson, Erik. 2003. *The Devil in the White City*. New York: Crown Publishers.

Layton, Edwin, Jr. 1986. *The Revolt of the Engineers: Social Responsibility and the American Engineering Profession*. Baltimore, Md.: The Johns Hopkins University Press.

Lewis, Oscar. 1938. *The Big Four*. New York: Alfred A. Knopf.

Lin, Tung-Yen, and Sidney Stotesbury. 1981. *Structural Concepts and Systems for Architects and Engineers*. New York: John Wiley & Sons.

Lord, C. C. 1890. *The History of Hopkinton*. Hopkinton, N.H.: privately published.

Lowry, Patricia. 2000. "George Washington Gale Ferris Slept Here . . . Didn't He?" *Pittsburgh Post-Gazette* (Aug. 26).

Macaulay, David. 2000. *Building Big*. New York: Houghton Mifflin Company.

Mansell, George. 1979. *Anatomy of Architecture*. New York: A and W Publishers.

Martin, Chlotilde. 1930. "Miss Kate Gleason," *The News and Courier* (Nov. 30), Charleston, S.C.

Maxwell, John. 1993. *Developing the Leader Within You*. Nashville, Tenn.: Thomas Nelson Publishers.

McCarrell, Stuart. 1996. *Voices, Insistent Voices: A Dramatization of Modern and Ancient Life in a Series of Strong yet Subtle Vignettes*. Chicago, Ill.: Xenia Press.

McCullough, David. 1972. *The Great Bridge*. New York: Simon & Schuster.

McCullough, David. 1977. *The Path Between the Seas*. New York: Simon & Schuster.

McDougall, Walter. 2004. *Freedom Just Around the Corner: American History 1585-1828*. New York: HarperCollins Publishers.

McNichol, Dan. 2003. *The Roads That Built America*. New York: Barnes and Noble/Barbara J. Morgan.

MSC. 1967. "Steel Solves Some Thorny Problems," *Modern Steel Construction* (3rd Quarter), American Institute of Steel Construction.

MT. 1935. "Marvel of Engineering and Work of Artistic Beauty," *The Morning Tribune* (Dec. 20), New Orleans, La..

Nautilus. 1988. "Remarkable Structures: America's Covered Bridges," *RWC Nautilus* (Spring), V. 22, N. 1, pp. 9–10, Denver, Colo.

Nautilus. 1984. "T.Y. Lin and His Peace Bridge," *RWC Nautilus* (Fall), V. 18, N. 3, pp. 4–6, Denver, Colo.

Nervi, Pier Luigi. 1956. *Structures*. New York: McGraw-Hill Book Company.

Noble, David. 1977. *America by Design: Science, Technology, and the Rise of Corporate Capitalism*. New York: Oxford University Press.

NB-ASCE. 2000. "Hardy Cross: Civil Engineer-Educator," (Biography compiled by the Norfolk Branch of ASCE), July 26, 2000.

Otto, Frei. 1973. *Tensile Structures*. Cambridge, Mass.: MIT Press.

PCG. 1893. "266 Feet in Air: The Ferris Wheel Turns and Mrs. Ferris Gives a Toast," *Pittsburgh Commercial Gazette* (June 17).

PCI. 2003. "Tung-Yen Lin (1912-2003)," *PCI Journal* (Nov.–Dec.), pp. 148–149.

Petroski, Henry. 1995. *Engineers of Dreams: Great Bridge Builders and the Spanning of America*. New York: Alfred a. Knopf, Inc.

Petroski, Henry. 1997. *Remaking the World: Adventures in Engineering*. New York: Alfred A. Knopf, Inc.

Plachta, Jan. 1997. "One Hundred Years of the Rock Island Government Bridge," *Journal of Bridge Engineering* (Nov.).

Plowden, David. 1974. *Bridges: The Spans of North America*. New York: W. W. Norton & Company.

Proffitt, Pamela (ed.). 1999. *Notable Women Scientists*. Farmington Hills, Mich.: Gale Group Inc.

Pursell, Carroll, Jr. (ed.). 1996. *Technology in America: A History of Individuals and Ideas*. Cambridge, Mass.: The MIT Press.

Ratigan, William. 1959. *Highways Over Broad Waters: Life and Times of David B. Steinman*. Grand Rapids, Mich.: William B. Eerdmans Publishing Co., Inc.

Reti, Ladislao (ed.). 1974. *The Unknown Leonardo*. New York: McGraw-Hill Book Company.

Reeve, W. Paul. 1995. "Even the Grasshoppers Were Starving During the 1934 Drought," *History Blazer* (Mar.).

Roehm, A. Wesley, et al. 1965. *The Record of Mankind*. Boston, Mass.: D.C. Heath and Company.

Robbins, Paul. 1984. *Building for Professional Growth: A History of the National Society of Professional Engineers 1934-1984*. Alexandria, Va.: NSPE Publication.

Russell, Jeffery (ed.). 2003. *Perspectives in Civil Engineering*. Reston, Va.: ASCE Press.

Salvadori, Mario, and Robert Heller. 1963. *Structure in Architecture*. Englewood Cliffs, N.J.: Prentice-Hall, Inc.

Salvadori, Mario. 1980. *Why Buildings Stand Up: The Strength of Architecture*. New York: W. W. Norton & Company.

Sandstrom, Gosta. 1970. *Man the Builder*. New York: McGraw-Hill Book Company.

Santi, Bruno. 1981. *Leonardo Da Vinci*. Florence, Italy: Scala Books (Distributed by Harper and Row, Publishers).

SAL. 1955. "Willard Eastman Simpson Man of Year," *San Antonio Light* (Feb. 21).

Schlager, Neil (ed.). 1994. *When Technology Fails*. Detroit, Mich.: Gale Research Publishers.

Scott, Stanley. 1994. *Connections: The EERI Oral History Series—Henry J. Degenkolb*. Oakland, Calif.: Earthquake Engineering Research Institute.

Sekine-Pettite, Cory. 1999. "Engineers of the Millennium." *CE News* (Dec.), pp. 64–66.

Severud, Fred, Jr. 2000. "Fred N. Severud: His Life," (White Paper Prepared for Richard Weingardt's Column in ASCE's *LME Journal*). Fred Severud Consulting Engineer, Salem, Mo.

Shapiro, Mary. 1983. *A Picture History of the Brooklyn Bridge*. New York: Dover Publications.

Shearer, Benjamin and Barbara Shearer (eds.). 1997. *Notable Women in the Physical Science*. Westport, Conn.: Greenwood Press.

Sheppard, Laurel. 1999. "Kate Gleason Leaves Her Legacy on RIT's College of Engineering," *SWE* (Sept./Oct.).

Sieden, Lloyd. 1989. *Buckminster Fuller's Universe: An Appreciation*. New York: Plenum Press.

Smith, Adam. 1976. *The Wealth of Nations*. Chicago, Ill.: The University of Chicago Press.

Smith, Hedrick. 1995. *Rethinking America*. New York: Random House.

Snyder, Carl. 1893. "Engineer Ferris and His Wheel," *The Review of Reviews: An International Magazine* (Jan.), Vol. VIII, pp. 268–276.

Snyder, David. 2001. "Demographic Realities in the Future of U.S. Engineering Companies," *Eye to the Future*. Washington, D.C.: ACEC Publications.

Sobel, Robert and John Raimo. 1979. *Biographical Directory of the Governors of the United States 1789-1978*, Westport, Conn.: Meckler Books.

Somerset. 1991. "Clifford Holland Remembered," *The Somerset Spectator* (Feb. 20), Somerset, Mass.

Somerset. 1974. "Clifford M. Holland, Great Tunnel Builder, Was a Somerset Boy," *The Somerset Spectator* (July 18), Somerset, Mass.

Steadman, H. Douglas. 1999. "The History of W. E. Simpson Company" (White Paper Furnished for Richard Weingardt's Column in ASCE's *LME Journal*), San Antonio, Tex.

Steinman, David. 1972. *Builders of the Bridge: The Story of John Roebling and His Son*. New York: Ayer Company Publishing.

Stone, Irving. 1956. *Men to Match My Mountains*. Garden City, N.Y.: Doubleday and Company.

Stuart, Charles. 1871. *Lives and Works of Civil and Military Engineers of America*. New York: D. Van Nostrand.

Swan, Russ. 1997. "TY Lin: Builder of Bridges," *Bridge Design and Engineering* (Nov.), Issue No. 9.

Tedesko, Anton. 1971. "Shells 1970—History and Outlook," *Concrete Thin Shells* (American Concrete Institute Publication SP-28), Detroit, Mich.

Temko, Allan. 1994. "Builder of Bridges: T. Y. Lin," *California Monthly* (Dec.), pp. 6–10.

Tiedeman, Jane. 2001. "John Greiner: A Brief Life History," (White Paper Prepared for Richard Weingardt's Column in ASCE's *LME Journal*). URS Corporation, Hunt Valley, Md.

Torroja, Eduardo. 1962. *Philosophy of Structures*. Berkeley, Calif.: University of California Press.

Trescott, Martha. 1990. "Women in the Intellectual Development of Engineering," *Women of Science*. Bloomington, Ind.: Indiana University Press.

Turner, Roland and Steven Goulden. 1981. *Great Engineers and Pioneers in Technology*. New York: St. Martin's Press.

Turak, Theodore. 1986. *William Le Baron Jenney: A Pioneer of Modern Architecture*. Ann Arbor, Mich.: UMI Research Press.

Vare, Ethlie Ann, and Greg Ptacek. 1987. *Mothers of Invention*. New York: Quill William Morrow.

Von Karman, Theodore. 2004. *Aerodynamics: Selected Topics in the Light of Their Historical Development*. Mineola, N.Y.: Dover Publications, Inc.

VP. 1959. "Dr. Hardy Cross: World Famed Engineer," *The Virginian-Pilot* (Feb. 12), pg. 12.

Watchtower. 1982. "From Atheism to a Purposeful Life," *The Watchtower* (Oct. 15), pp. 12–15.

Weigold, Marilyn. 1984. *Silent Builder: Emily Warren Roebling and the Brooklyn Bridge*. Port Washington, N.Y.: Associated Faculty Press, Inc.

Weingardt, Richard. 2004. "Engineering History, Travel and You," *Structural Engineer* (June), pg. 14.

Weingardt, Richard. 1998. *Forks in the Road: Impacting the World Around Us*. Denver, Colo.: Palamar Press.

Weingardt, Richard. 1999. *RAUT: Teacher-Leader-Engineer*. Boulder, Colo.: University of Colorado College of Engineering Publishing.

Weingardt, Richard. 2003a. "Structural Engineering Legends," *Structural Engineer* (Sept.), pg. 16.

Weingardt, Richard. 2003b. "Tomorrow's Engineer Must Run Things, Not Just Make Them Run," *Perspectives in Civil Engineering*, pp. 385–398. Reston, Va.: ASCE Press.

Whitford, Noble. 1906. *History of the Canal System of the State of New York*. Albany, N.Y.: Brandow Printing Company.

Widegren, Ragnar. 1988. *Consulting Engineers 1913–1988: FIDIC Over 75 Years*. Stockholm, Sweden: AB Grafiska Gruppen.

Wood, Paul. 2003. "University of Illinois to Get Papers of Ex-NASA Engineer," *The News-Gazette*, Oct. 9.

WSE. 1992. "From the Archives—A Personal Reminiscence of Octave Chanute by Warren R. Roberts," *Midwest Engineer*. Chicago, Ill.: Western Society of Engineers.

WSE. 1970. *The Centennial of the Engineer*. Chicago, Ill.: Western Society of Engineers.

Yang, Sarah. 2003. "T.Y. Lin, World Renowned Structural Engineer, Dies at Age 91," *UC Berkeley News* (Nov. 18).

Young, David. 1996. "The Wright Stuff" (Contributions of Octave Chanute), *Chicago Tribune*, June 23, Section 12, pp. 4–5.

Zahm, Albert. 1911. "Octave Chanute, His Work and Influence in Aeronautics," *Scientific America*, May, pp. 463–488.

Zils, John. 2000. "Fazlur Rahman Khan: A Brief History," (White Paper Prepared for Richard Weingardt's Column in ASCE's *LME Journal*). SOM, Chicago, Ill.

Index

About the Author

Richard George Weingardt, a third-generation native of Colorado, is the author of eight books and some 600 published papers and articles on engineering, business, leadership, and creativity. In 2003, his book *Forks in the Road* won the American Association of Engineering Societies (AAES) Journalism Award. Weingardt is the first and only engineer to have received the award. A motivational speaker and practitioner, he has been active in both professional and community groups. He was the 1995–1996 national president of the American Council of Engineering Companies (ACEC) and the 2002–2003 chair of ASCE's Communication Committee (CCOM).

A graduate of the University of Colorado with master's and bachelor's degrees in civil-structural engineering, Weingardt is a registered professional engineer in 28 states. The consulting engineering firm he founded in 1966 has completed more than 4,500 major projects worldwide. Many of them have received engineering excellent awards, among them, the three airside terminals—Concourses A, B, and C—at Denver International Airport, Jefferson County Courthouse, Integrated Teaching and Learning, and Discovery Learning Laboratories at the University of Colorado.

Weingardt and his wife, Evie, reside in Denver, Colorado.